Digital Intelligent Aeroengine

Quanyong Xu · Jiali Yang · Wenyu Cao ·
Zhongzhi Hu · Ai He · Kai Liu

Digital Intelligent Aeroengine

Revolutionary Technologies for Aeroengine

Springer

Quanyong Xu
Institute for Aero Engine
Tsinghua University
Beijing, China

Jiali Yang
Institute for Aero Engine
Tsinghua University
Beijing, China

Wenyu Cao
Institute for Aero Engine
Tsinghua University
Beijing, China

Zhongzhi Hu
Institute for Aero Engine
Tsinghua University
Beijing, China

Ai He
Institute for Aero Engine
Tsinghua University
Beijing, China

Kai Liu
Institute for Aero Engine
Tsinghua University
Beijing, China

ISBN 978-981-95-0941-6 ISBN 978-981-95-0942-3 (eBook)
https://doi.org/10.1007/978-981-95-0942-3

This work was supported by Tsinghua University (20234616001).

This Springer imprint is published by the registered company Springer Nature Singapore Pte Ltd.
The registered company address is: 152 Beach Road, #21-01/04 Gateway East, Singapore 189721, Singapore

Acknowledgements

Special thanks are extended to Zhu Yuqing, author of the chapter "Data-Driven Intelligence Empowers Next-Generation Aircraft Engine Innovation", and to Yang Shubo, author of the chapter "Auxiliary Generation Technology for Engine Model Code on the Basis of Prompt Engineering". Their contributions have significantly enhanced the impact of this book.

List of Chapter Authors

Title: Current Status, Challenges and Key Issues of Digital Engine Technology
Authors: Hu Zhongzhi[1], Cao Wenyu[1], He Ai[1], Yu Jianwei[1], Liu Kai[1], Yang Jiali[1], and Xu Quanyong[1]

Title: Simulation Modeling Method for Aero Engines Based on Digital Twins
Author: Xu Quanyong[1]

Title: Enhanced Model of a Whole Aircraft Engine System
Author: Yang Jiali[1]

Title: Research on a Construction Method for Digital Thread Based on a Central Semantic Model
Author: Cao Wenyu[1]

Title: Cross-dimensional Model on Overall Performance Calculation Method of the Turbofan
Author: Xu Quanyong[1]

Title: Nonlinear Identification Method for Aircraft Engine Component Health Parameters Based on Proximate Operating Conditions
Author: Liu Kai[1]

Title: Auxiliary Generation Technology for Engine Model Code on the Basis of Prompt Engineering
Author: Yang Shubo[1]

Title: AI-Based Digital Maintenance Technology for Aircraft Engines
Author: He Ai[1]

Title: Data-Driven Intelligence Empowers Next-Generation Aircraft Engine Innovation
Author: Zhu Yuqing[2]

[1]Tsinghua University, Institute for Aero Engine, Tsinghua Garden, Haidian District, Beijing, China
[2]Tsinghua University Beijing National Research Center for Information Science and Technology, Tsinghua Garden, Haidian District, Beijing, China

Preface

With the rapid development of artificial intelligence, big data and cloud computing technology, the core power device of aviation, the engine, is facing a key period of digital and intelligent transformation. This book systematically reviews the latest progress, core challenges and future directions of digital engine technology, with the aim of providing engineering technicians, scientific researchers and industry decision-makers with an authoritative reference that integrates multidisciplinary perspectives and discusses technological innovation and application practices.

The book opens with a focus on the current status and challenges of digital engine technology and reveals its enormous potential for improving design efficiency, optimizing performance and reducing maintenance costs while highlighting that data integration, model accuracy and real-time performance are still the key bottlenecks that restrict the application of this technology. In response to these problems, subsequent chapters explore innovative solutions from different dimensions.

Chapters "Simulation Modeling Method for Aero Engines Based on Digital Twins"–"Research on a Construction Method for Digital Thread Based on a Central Semantic Model" delve into simulation modeling methods that are based on digital twins and digital mainlines. The "mechanism-driven + data-driven" collaborative strategy is proposed, which involves correcting physical models through machine learning to achieve high-fidelity digital twins, overcoming the limitations of traditional single-discipline simulation, constructing enhanced whole-machine system models, and verifying the rationality of their designs through multisystem joint simulation. More innovatively, digital mainline technology that is based on a central semantic model is proposed, which uses unified metadata standards to break through data silos throughout the lifecycle, thereby supporting rapid iteration and quality improvement.

Chapters "Cross-dimensional Model on Overall Performance Calculation Method of the Turbofan"–"AI-Based Digital Maintenance Technology for Aircraft Engines" focus on multiscale simulation and intelligent diagnostic technology. By further expanding the cross-dimensional simulation framework, a method of interpretation that integrates mechanisms and data, overcomes the dilemmas of traditional models, and quantitatively evaluates the credibility of simulations is proposed. Furthermore, prompt engineering is applied to model code generation to significantly improve development efficiency, and an AI-driven fault diagnosis system that achieves real-time and accurate engine health management through deep learning is systematically described.

Chapter "Data-Driven Intelligence Empowers Next-Generation Aircraft Engine Innovation" examines the future ecosystem of digital intelligent engines from a global perspective and proposes a "data-intelligence" dual-drive innovation model. This chapter emphasizes how the deep integration of big data and AI algorithms will reshape the engine R&D system and demonstrates the path of technology implementation through typical cases.

The greatest feature of this book, which includes both basic theoretical breakthroughs and innovative engineering applications, is its interdisciplinary integration and practice-oriented nature. It not only provides a method reference for professional researchers but also provides a feasible technical route for the implementation of intelligent transformation in industry. This study provides not only a systematic summary of achievements in the field of aviation engine digital technology in our country in recent years but also an exploration declaration for the future intelligent power revolution. We hope that this book can contribute to promoting the digital transformation of aviation engines in our country.

Beijing, China

Quanyong Xu
Jiali Yang
Wenyu Cao
Zhongzhi Hu
Ai He
Kai Liu

Competing Interests The authors have no competing interests to declare that are relevant to the content of this manuscript.

Contents

Current Status, Challenges and Key Issues of Digital Engine Technology

Abstract Aero engine design simulation technology is gradually developing toward integration and coupling, accompanied by improvements in system engineering methodology, and the field of aero engines is currently undergoing a digital transformation. This article reviews the concepts, current status, needs, and key technical issues of aero engine digitalization, with the aim of providing a reference and support for the digital transformation of aero engines. Given the complexity of aero engines, the concepts and substantive connotations of model-based system engineering, digital twins, and digital threads are analyzed, and a definition of digital engines is developed. The use cases and development status of digital engines in China and other countries are summarized, and the work that needs to be carried out to address technical trends and challenges, such as model algorithm libraries, general platforms, data connectivity and processing, coupled integrated simulation technology, data analysis and decision-making, is discussed. The content and directions of research in China and other countries on the key technical issues of digital transformation, such as discipline simulation integration methods, system model and discipline model connection methods, and digital twin modeling methods, are discussed. The ideas and specific research content for digital engines can provide a basis for subsequent planning and construction.

Keywords Aero engine · Digital engineering · Digital thread · Digital twin · Model-based system engineering

Introduction

Aero engines are multidisciplinary integrated power devices that need to meet many stringent and contradictory comprehensive index requirements and operate in harsh working environments for a long time, which poses great challenges to technology research and development and engineering design. To meet the increasing performance requirements and technological innovation requirements of aircraft, new aero engine configurations and technical solutions are constantly emerging, the research

Q. Xu et al., *Digital Intelligent Aeroengine*,
https://doi.org/10.1007/978-981-95-0942-3_1

and development cycle length and costs are increasing, and the entire aero engine research and development process faces considerable challenges.

The aero engine research and development process is a full life cycle process in which the concept of system engineering is implemented, and related specifications have been formulated, such as the SAE-ARP4754 "A Civil Aircraft and System Development Guide in the United States" and the "Civil Aero Engine Product Development Stage Management Detailed Rules (Trial)" in China. The full life cycle of an aero engine product is divided into a requirement analysis and definition stage, a design stage, a manufacturing stage, and an operation and maintenance stage. Owing to the collaborative and multidisciplinary nature of aero engine development, as well as the need to integrate many components and related models, the promotion of the digital process and the implementation and use of advanced system engineering methods and tools to effectively manage product complexity are urgently needed.

Simulation has become an important part of the research and development process. In the design and development stage, the manufacturing implementation stage, and the integration verification stage, the introduction of simulation technology has become a key factor in shortening the development cycle of aero engines [1]. With the further development of simulation technology and digital technology, aero engine design simulation technology is gradually developing toward integration and fusion. Moreover, with further improvements in system engineering methodology, the entire aero engine research and development system is undergoing revolutionary changes. All aero engine stakeholders are eagerly looking for solutions related to model-based system engineering (MBSE), digital twins, and digital threads.

This article focuses on the digital development of the aero engine field, proposes the concept of digital engines, and analyzes the current status, challenges, needs, and key technical issues of aero engine digital development to help industry catch up with the state of the art in digital and intelligent development and establish an new aero engine research and development system, thus providing an opportunity for transformation.

Digital Engine Definition

Owing to the high requirements for product performance and the strict control of development risks and costs, the aviation industry has always been the source and pioneer of digital transformation. Compared with aircraft products, aero engines have higher complexity, which is manifested mainly in the interactions among many subsystems and components, which need to be defined, specified, and verified throughout the life cycle. Moreover, the dynamic interactions among various parts of an aero engine and the dynamic characteristics under various operating conditions, especially the coupling of multiple components and multiple simultaneous physical phenomena, lead to great challenges in accurately and completely predicting these interactions and characteristics from physical mechanisms [2]. Considering the complexity of aero engines, the large amounts of data created in different ways must be handled and managed in a consistent manner throughout the life cycle of aero engines, and more detailed expressions and descriptions of different levels of design, manufacturing, and operation and maintenance granularity, as well as various procedures for system and subsystem design and integration, can be effectively managed only by adopting system engineering processes and methods to address these multidomain, multiparticipant, and multilevel issues [3].

Although the industry is making continuous advancements in data management systems on the basis of system engineering and the full life cycle process on the basis of simulation, owing to the complexity of aero engines, the current research and design (R&D) system cannot meet the development requirements of future aero engines. Large amounts of information on concepts, knowledge, behavior, intentions, logic, and relationships are not reflected or controlled in this system; the association, integration, and traceability of multisource data and cross-domain models are still lacking in terms of management; standardized modeling, driving and reuse processes have not been established; research on multidisciplinary coupled integrated simulation technology oriented to requirements needs to be strengthened; and the integrated simulation and decision-making capabilities based on data and intelligence are not sufficient. A large amount of uncertainty continues to enter each product's lifecycle, and R&D personnel are consumed with procedural and transactional work and are unable to focus on innovation [4]. On this basis, new interrelated concepts and methodologies are constantly being proposed to promote the digital paradigm shift from different angles.

With the rapid development of computer and information technology and engineering technology in various fields, the technology of describing systems with object-oriented, visual system modeling languages is gradually maturing. MBSE supports system requirements and design, analysis, verification and confirmation activities with standardized system modeling technology; runs continuously through the development of complex products; can effectively solve the problems faced by system engineering on the basis of descriptive text information in data interaction and technical status management; and can improve the consistency of system full-cycle information representation through language and tools [5]. MBSE does not focus on solving design problems in a specific discipline. System modeling language actually describes "information about data". It is a tool for managing data, models, repositories and simulations and can also be used to define data interoperability, exchange, integration, linking, traceability, reuse, conversion, archiving and synchronization. MBSE emphasizes modeling oriented to the system engineering process; modeling and forming organic connections among analysis elements involved in system requirements, system analysis, system design, system verification and other processes; and maintaining the consistency and traceability of system information throughout the lifecycle [6].

A digital thread provides a data communication architecture with shareable resources (such as sensor outputs, computational tools, methods and processes), links information from all stages of the product lifecycle (such as early concepts, design, manufacturing, operation, postlife and retirement) to achieve real-time and long-term decision-making, and can also be considered a product's main or "authoritative" data and communication platform [7]. From a data perspective, the means of realization include standardized data definitions, authoritative semantic representations of systems, and data cleaning and mining. From the perspective of design tool integration, the methods for realization include data linking, model-driven and code-based automatically generation, model and data exchange on the basis of standards (for example, FMI and STEP), and conversion on the basis of "meta-models" [8].

Aero engines have been simulated numerically in almost every field and at every system level to explore the design space in detail and achieve significant performance improvements. The key to aero engine simulation is the integration and coupling of multilevel, multidomain structures and behaviors to accurately describe system characteristics. The concept of digital twins aims to achieve high-precision expressions of aero engines, which can be virtually verified before entering the physical processing stage to ensure correct design decisions, improve the credibility of simulation and test data, reduce the number of experiments, reduce the number of onsite failures, make real-time decisions, and apply high-precision models, actual usage data and historical data to simulate real-world behavior under all conditions and environments [9, 10]. High-precision models are not necessarily based on physical mechanisms, nor do they always have higher dimensions or smaller scales. More consideration is given to meeting the accuracy, resolution, real-time performance and economic requirements [11]. In current practice, most digital twin models focus on reducing model uncertainty or making decisions through data fusion.

The above concepts and their related methodologies and tools are generally applicable to industrial products; however, for equipment R&D systems, problems cannot be considered in isolation. In 2018, the U.S. Department of Defense officially released the "Department of Defense Digital Engineering Strategy", which integrates related concepts and methodologies from a higher digital dimension; covers the technical and management processes of system engineering, with system models, digital threads and digital twins that run through the equipment system as the core; and builds a digital engineering ecosystem centered on digital models [12]. Although digital engineering targets the entire equipment industrial system, the specific practice processes for different products differ significantly. The core aspect of digital transformation is the digitization of expert knowledge, system expression, and behavioral intent.

In product lifecycle management, digital mock-ups are widely used. A digital mock-up is a digital description of a whole machine or subsystem with independent functions of mechanical products. This description reflects not only the geometric properties of the product object but also the function and performance of the product object in at least one field (GB/T 26,100–2010). Although a generalized digital prototype also includes functional and performance descriptions of the product, it does not emphasize the completeness or unity of each version. In response to the new digitalization trend in the field of aero engines, we further develop the concept of digital prototypes and propose digital engines. A digital engine is defined as a cross-level, cross-scale, multidisciplinary model with data integration based on a unified and complete engine system modeling expression that covers all aspects of the engine

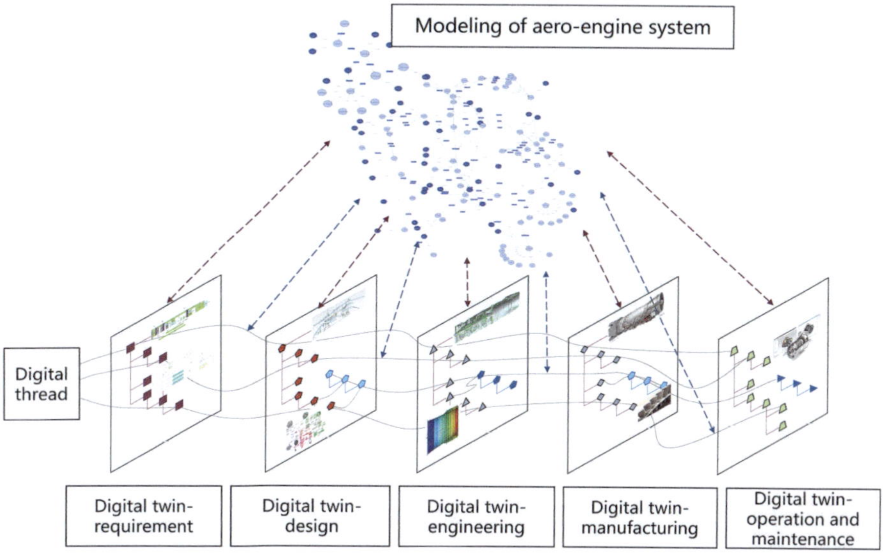

Fig. 1 Schematic diagram of the digital engine concept

life cycle, thus ensuring the traceability and consistency of data, achieving high-precision and high-confidence representation, summarizing and integrating engine system functions and behaviors at different levels, and providing support for decision-making (see Fig. 1). A digital platform that can support the development of digital engines is called a digital engine platform.

Current Developments

In response to the developments toward digitalization and intelligence in the field of aero engines, relevant activities have been carried out in China and other countries. Digitalization runs through the entire life cycle of aero engine design and development and changes the design, manufacturing, integration, verification, operation and maintenance methods for aero engine products. At present, model-based system engineering methods and tools are gradually being applied by mainstream aero engine manufacturers, and unified collaborative platforms and data management architectures with digital mainline connectivity have gradually been developed. The updating of existing simulation processes and model libraries to form multilevel and multiview integrated simulation capabilities initially achieved promising results. The importance of data-based models and intelligent decision-making capabilities is gaining widespread consensus. The applications of and research directions for digital engines are listed in Table 1.

Table 1 Summary of digital engine applications and cases

Company	Applications	Direction	Evaluation	Sources
General electric	Researching high-fidelity MBSE technology and tool chains; formulating MBSE digital engineering strategies; developing related technologies, tools and processes to improve safety, resilience and cycle time; developing system tools to promote integration; using open-source tools; and shortening the process cycle by 50% and the verification and acceptance times to 1/10 [13]	MBSE	The related work has just begun, and the whole idea is consistent with the US military's digital engineering, which is promoted by the R&D center; however, the technical maturity needs to be improved	
General electric	The Predix platform is an industrial internet platform and learning system that is optimized for digital twins and supports industrial big data analysis on the basis of digital twins	Digital twins	Due to the rich expert knowledge of General Electric, the Predix platform has been successfully applied in aero engines and gas turbines, but in the commercialization process, customers do not have good data or model accumulation, and there are various problems in promotion	https://www.ge.com/cn/b2b/digital/predix
General electric	Digital threads for design can quickly create artificial intelligence agent models according to engineering workflows and simultaneously coordinate and thread related data. This project focuses on applying capabilities to each use case of design, model, and simulation to enable rapid deployment and search and on integration with existing product lifecycle database systems to automate workflows and agent models	Digital threads	Emphasizes the use of machine learning agent models, which can be quickly or even real-time deployed through mechanism models and large amounts of training on accumulated data to realize rapid iteration and decision-making	https://www.ge.com/research/technologydomains/digital-technologies/digital-thread-design

(continued)

Table 1 (continued)

Company	Applications	Direction	Evaluation	Sources
General electric	Using sensors and big data analysis to improve the speed and efficiency of manufacturing and to connect design and manufacturing objectives via digital thread design. A design is transmitted for manufacturing engineering in a digital way, the process is modeled and simulated virtually, and the factory process and layout and manufacturing control are also simulated and optimized. Once the design and process are verified virtually, the data are transmitted to a smart factory, where they are converted into parts or products, to achieve optimal production control and logistics through real-time digital connection	Digital threads	Bridging the gap between design and manufacturing, achieving integrated design-process analysis and full-process simulation, and especially optimizing manufacturing strategies and costs can yield significant benefits	https://manufacturing.cioreview.com/cxoinsight/digital-threadcreating-a-self%20improving-brilliant%20factory-nid-194-cid-34.html
General electric	Combining MBSE with feature-based product line engineering to verify digital structure requirements and test digital structure language functions and automatically generating T901 engine controller software	MBSE	The association of MBSE models and discipline models formed by decomposing requirements is the core aspect, and a competitive advantage has been achieved in this direction	https://biglever.com/momentumblog-series4/

(continued)

Table 1 (continued)

Company	Applications	Direction	Evaluation	Sources
General electric	Collecting data from jet engines and turboprop engines and combining these data with physical models to predict faults and frequencies, creation of a digital twin on the basis of performance models, monitoring engine performance, predicting maintenance issues, and reducing costs throughout the life cycle. The engines include the GEnx, GE90, CF6, CF34, CFM56, and LEAP engines	Digital twins	Digital twins are currently the most mainstream application. The uncertainty of the mechanism model is reduced via data collection, and the maintenance cost of aero engines is also reduced, but this technology depends on the accuracy of the engine physical model	https://blog.geaerospace.com/technology/no-two-alike-each-jet-engine-delivery-comes-with-its-own-digital-thumbprint/
Pratt & Whitney	Adopting MBSE and digital thread methods, standards, and tools; proposing a platform architecture that enables the use of digital twins through digital threads; emphasizing standardized semantics and open service architectures to solve tool interoperability and integration problems; and using semantic integration and open service architectures to bridge digital twins of functional islands	Digital threads	Proposed a holistic digital architecture for aero engines, which inherits the properties of the existing PLM system and is in line with the digital engineering thinking, and proposed the importance of semantic integration and open service architectures. The architecture can be used as a reference for applications in China	
Pratt & Whitney	Conducting advanced diagnostics and engine monitoring services through the EngineWise platform, using networked software tools, and providing health data analysis for more than 8000 engines in use. The company and its engine operators can use predictive AI to propose thrust suggestions	Digital twins The same platform functionality as that of General Electric and the most easily profitable application direction	The same platform functionality as that of General Electric and the most easily profitable application direction	https://prattwhitney.com/products-and-services/services/enginewise

(continued)

Table 1 (continued)

Company	Applications	Direction	Evaluation	Sources
Pratt & Whitney	Promoting the Connected Factory pilot project, building a digital thread to drive higher levels of process capabilities, optimizing with lean principles, achieving automation where appropriate, collecting data in a connected environment, running analysis, and providing real-time information to operators and leaders	Digital threads	Further development under the existing system, cautiously advancing, reducing costs in the production process, but the deployment cost is high	https://newsroom.prattwhitney.com/Pratt-Whitney-Launches-Connected-Factory-Initiative
Pratt & Whitney	For the in-service F119 engine, establishing its digital twin, predicting a 20% increase in component life, and saving $800 million in costs or extracting more engine performance by delaying the replacement of certain components during the aircraft's life cycle. Completed the DEEC software update using agile software development methods, including regression testing and 100-h verification, and improving the performance in specific areas within the F-22 flight envelope	Digital twins	A typical use case of digital twins for military engines. In addition to achieving life extension for existing engines, it can also balance service life and high performance	https://aviationweek.com/defense-space/aircraft-propulsion/digital-twin-helps-pratt-expand-f119-performance-f-22
Rolls–Royce	Proposed the Intelligent Engine platform, which establishes a two-way information flow to obtain the engine operating environment and provides customized operation and maintenance suggestions for each engine. Intelligent decision-making and self-maintenance are supported through artificial intelligence and advanced analytics	Digital twins	All three major civil engine companies provide related services, which are also typical applications of digital twins	https://www.rolls-royce.com/media/our-stories/discover/201/intelligent-engine-health-monitoring.aspx

(continued)

Table 1 (continued)

Company	Applications	Direction	Evaluation	Sources
Rolls–Royce	To win the B-52 re-engineering contract, Rolls–Royce established a digital twin model of the F130 engine by integrating the engine digitally into pod, pylon, and wing components and dynamically simulating the entire operating system. This enables customers to accurately simulate the efficiency of the product on the wing and demonstrates maintenance work	Digital twins	Visualization is performed for digital twins, which enables customers to intuitively understand the entire re-engining process and risks. Visualization is an important development direction for digital twins	https://www.airandspaceforces.com/rolls-royce-digitally-modeled-entire-wing-pylon-to-win-b-52-engine-contract/
Rolls–Royce	Rolls–Royce uses MBSE/SysML to obtain full authority control system and software level requirements to form a SysML symbol subset and model control system modeling standards. It promotes the execution of automatic verification scripts for system- and software-level models to ensure consistency, correctness (where possible), and compliance with modeling standards [14]	MBSE	Concerned with the interaction standards of MBSE requirement decomposition and control system discipline models. It has been piloted in some subsystems of aero engines. The maturity of the technology needs to be improved, and the use of open-source software is beneficial for interaction and collaboration	

(continued)

Table 1 (continued)

Company	Applications	Direction	Evaluation	Sources
Rolls–Royce	For the Ultarfan engine power gearbox, MBSE modeling was carried out for the lubrication system, with requirements and definitions integrated into the model, thus achieving safety and modeling integration (in compliance with ARP4754A standards). A joint model that covers the complete product system hierarchy was created. Each individual model in the joint model defines a system architecture based on functional and spatial constraints, fault propagation, and emergent behavior, which enables this system hierarchy to be developed in parallel and to support iteratio	MBSE	The entire system modeling process of an engine system has been fully implemented, and the MBSE method has entered the substantive deployment stage and is gradually entering the new product development process	https://www.eclipse.org/capella/resources/pdf/Case_Study_Rolls_Royce.pdf
Safran	Safran established a design–analysis integrated platform for the entire lifecycle from preliminary design to detailed verification and acceptance via behavioral digital engine technology. This platform develops standards for data sharing and exchange behavior in heterogeneous environments and integrates applications of multiple information systems, thereby achieving data integration [2]	Digital threads	Focusing on system modeling, configuration management, and design–simulation integration issues, it has achieved full-process digital thread technology based on a certain type of aero engine, with a high degree of completion	
Safran	For turbofan engines, Safran built a small-scale digital twin solution for analysis. Using new cluster technology and adjustment algorithms, it reduces the observation dimensions for mathematical calculations. It uses autoencoder neural networks and cluster analysis for parallel deployment on computer clusters [15]	Digital twins	This was a demonstration project, the maturity of the technology needs to be further improved, and the algorithm has certain reference significance	

(continued)

Table 1 (continued)

Company	Applications	Direction	Evaluation	Sources
UEC	UEC proposed a digital twin architecture on the basis of a digital platform. Considering the requirement matrix, it introduces different degrees of complexity into the digital model, which can complete certification and predict engine operation. Pilot projects were implemented on the TV7-117ST engine and the AI222-25 engine [16]	Digital twins	A complete digital twin architecture was proposed on top of the existing design system, and pilot applications were carried out on the basis of real models, but the actual effect needs further verification	
AG PAD	AG PAD promotes the identification and description of the current capabilities of MBSE standards and tools in supporting industry data exchange. Digital twin/digital thread business architecture and methods are discussed; a comparative analysis of standards, strategies and roadmaps is conducted; and specific implementation measures, such as multiview BOM and definitions based on 3D models, are explored. Through digital tools and model-based processes, collaboration in the aviation industry is achieved through two-way exchange of technical data packages. The concept design process of model-based systems engineering (MBSE) is implemented to evaluate the exchange of digital requirements and system architecture models in collaborative product development activities	MBSE	The greatest significance of this project is the unanimous confirmation that data exchange, business systems, and implementation methods must be led and defined by the industry, not by software vendors	https://www.cimdata.com/en/aerospace-and-defense

On the basis of a wealth of historical data and strong R&D experience, General Electric began the process of R&D digitalization, which was marked by PDM and simulation process construction, in the 1990s. This process gradually formed a global collaborative design system based on processes and applications, with the integration of nearly 2000 applications, including business, enabling, and engineering. The main achievements were as follows: reducing cycle time through increased collaboration between engineering and supply chain functions; development of a universal user interface, which ensures a unified experience and shared digital reference; ensuring that each stakeholder views the latest information in the relevant context; and optimizing customer demand response time. The basic aims of data-centricity and intuitive search functions are improving productivity, simplifying data management, and promoting efficient cross-border information sharing. On the basis of a mature digital design system, General Electric's engine R&D system is undergoing vigorous renewal to further shorten the design and certification cycle through high-fidelity MBSE technology and tool chain engineering [13].

General Electric pioneered the industrial internet and launched the Predix platform, on which related work on digital engines has been carried out. The platform provides information transmission capabilities centered on industrial equipment data mapping and analysis to provide strong support for industrial application APPs. A multilevel cloud architecture was developed from industrial internet infrastructure services and platforms to software and services, which relies on the digital mainline of the entire lifecycle. All data models can communicate bidirectionally, so the status and parameters of real physical products provide feedback to the digital model through the physical system integrated with the intelligent production system, which keeps the digital model of each link of the lifecycle consistent, thus achieving dynamic, real-time evaluation of the current and future functions and performance of the system. Through the digital model, the platform can debug, test and optimize its operating state in a virtual environment, thus forming a digital twin for specific individual products. During operation and maintenance, a digital twin model is established for the on-wing engine, a fleet monitoring, diagnosis and health management framework is constructed, and fault isolation, fault detection and quantification functions are achieved [17, 18].

On the basis of many professional models, the digital thread for design (DT4D) is a capability that General Electric is developing, which can quickly create artificial intelligence agent models according to workflows while coordinating and threading model-related data. This project focuses on applying DT4D capabilities to each use case of design, modeling, and simulation to enable rapid deployment and search while integrating with existing product lifecycle management (PLM) systems to achieve automation of workflows and agent models (see Fig. 2).

Based on a mature R&D system, Pratt & Whitney used digital engineering to connect different tools and optimize automation, thus realizing integration across multiple domains through a digital thread, while adopting model-based system engineering to collaborate with upstream and downstream customers and suppliers, thereby successfully promoting system innovation and forming the architecture of the digital engine platform [19] (see Fig. 3). Facing the gap between design and manufacturing information, Pratt & Whitney advanced the Connected Factory project to further implement the landing of the digital thread.

The Engine Wise platform was created in 2017 and includes the Advanced Diagnostics and Engine Monitoring System (ADEM), which is a network-supported software tool with engine health management services that provides health data analysis for more than 8000 engines in use. The company and its engine operators can use

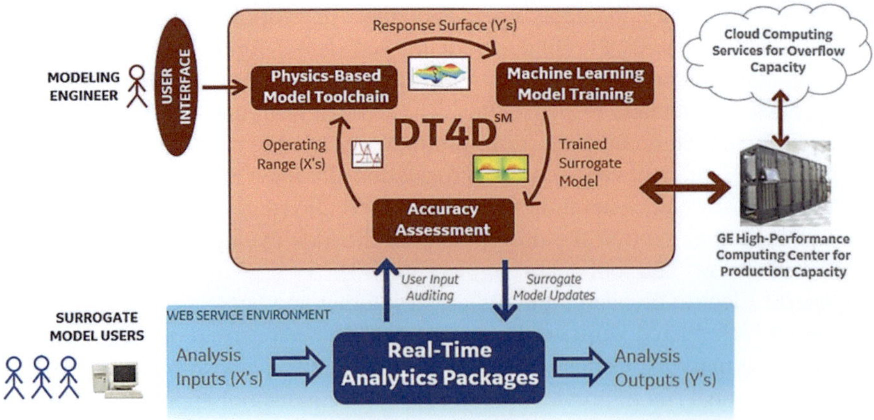

Fig. 2 General Electric's digital thread for design and R&D

artificial intelligence to predict and propose recommendations for increasing fuel economy by 15% and reducing operational interruptions and downtime. For the in-service F119, Pratt & Whitney established its digital twin, predicted a 20% increase in component life, and completed DEEC software updates using agile software development methods, including regression testing and 100-h verification, which improved the performance of specific areas of the F-22 flight envelope.

Rolls–Royce developed and perfected a solution for the transmission of product data up and down the digital mainline by obtaining engineering design and definition data during the manufacturing and service stages, with the aim of enabling all users to see the same state of a single design at any point. MBSE modeling was carried out for the lubrication system of the UltraFan engine power gearbox by integrating requirements and definitions into the model, defining the system architecture, and achieving safety and modeling integration (in compliance with the ARP4754A standards). To capture all the interactions, required functions, emergency behaviors, systems and components of all the oil systems, Rolls–Royce's solution involved creating a joint model that covers the complete product system hierarchy. Each individual model in the joint model defines a constraint system architecture on the basis of functional and spatial constraints, fault propagation, and sudden behaviors, which enables parallel development of this system hierarchy and promotes iterative system development. The company also focused on adopting advanced data analysis, industrial artificial intelligence, and machine learning to accelerate the development of new data analysis and services, obtain new value from the data, combine the latest analysis techniques with industry knowledge and engineering expertise, and better support customers. Rolls–Royce generated a large amount of data through engine operations, conducted extensive data analysis, and then created a digital twin of the engine to provide a richer dataset. Rolls–Royce then built and tested a model, made decisions, and then delivered the results to users through expert knowledge and data analysis methods. The

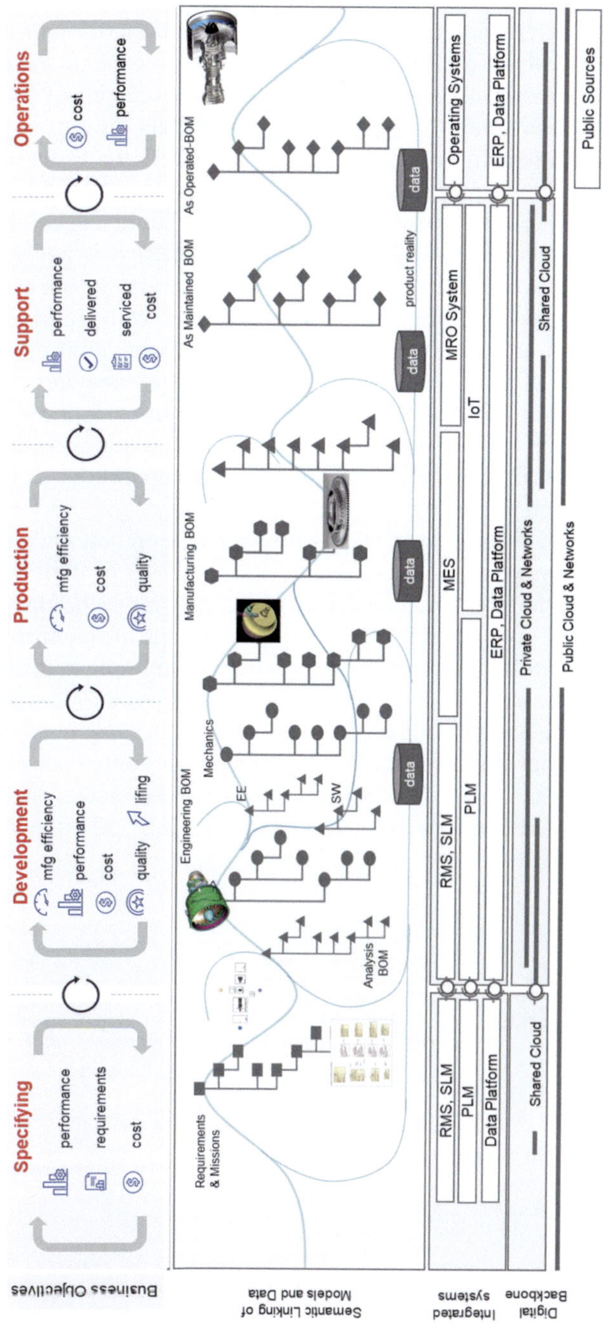

Fig. 3 Pratt & Whitney's aviation engine digital platform architecture [19]

digital twin is part of Rolls–Royce's "Intelligent Engine" vision, which seamlessly combines the physical engine with maintenance services and digital technology.

To strengthen the integration and collaborative capabilities of multidisciplinary system integration and whole-machine digital modeling in enterprises, reduce dependence on physical experiments, and accelerate product development processes, in 2009, funded by the European Commission, led by Airbus, and in collaboration with 59 partners and research institutions from 13 countries, the Collaborative and Robust Engineering using Simulation Capability Enabling Next Design Optimization (CRESCENDO) project and subsequent plans were launched on the basis of the concept of simulation-driven design optimization. Much work has been carried out on the aero engine, which is its most important subsystem. Safran established a design–analysis integrated platform for the whole life cycle, from preliminary design to detailed verification and acceptance, on the basis of behavioral digital engine technology; developed norms for data sharing and exchange behaviors in heterogeneous environments; the integrated application of multiple information systems; and traced and audited the input and output and method implementation of each design link in the design process, thus establishing collaborative capabilities on the basis of simulation-driven design [2].

The United Engine Corporation (UEC) of Russia has been committed to formulating an engine life cycle management strategy on the basis of digital twin technology. They have laid the foundation for the use of digital twins as the main tools. The target parameters of the product during the development stage are confirmed through a requirement management system. Pilot projects have been implemented on UEC's TV7-117ST engine, Cannon Company's AI222-25 engine, and Saturn Company's marine gas turbine. The Saturn Company developed a product requirement management module, carried out a large amount of high-precision mathematical modeling of marine engine elements, created a whole-machine dynamic model, and developed related programs for predicting operating states and fault trend development [16], as shown in Fig. 4.

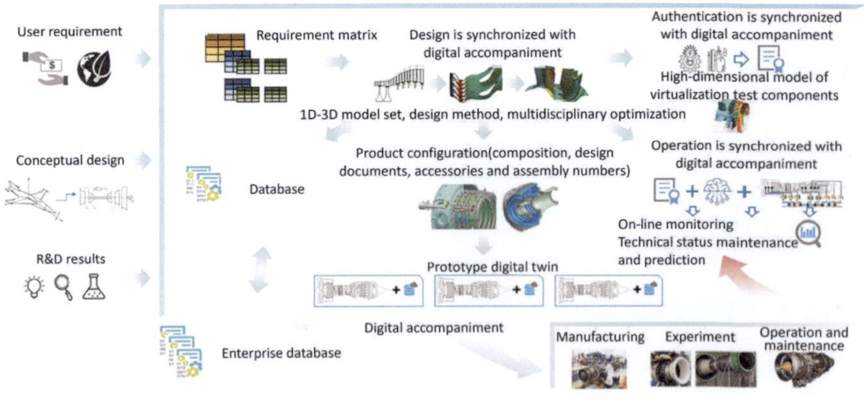

Fig. 4 UEC's aviation engine digital twin platform architecture [16]

The Aerospace and Defense Product Lifecycle Management Action Group (AD PAG) is an association organization of aerospace original equipment manufacturers and aircraft engine suppliers, including Airbus, Boeing, Embraer, General, Gulfstream, Mitsubishi Regional Jet, Pratt & Whitney, Rolls–Royce, and Safran. It identifies and describes the current capabilities of MBSE standards and tools in supporting industry data exchange; discusses digital business system architecture and methods; performs comparative analyses of standards, strategies and roadmaps; and explores specific implementation measures such as multiview BOM and 3D model-based definitions. The group's main business challenges include digital tools and model-based processes, which achieve industry collaboration through two-way exchange of technical data packages. Project teams composed of domain experts from member companies have been established to evaluate current data interoperability standards, implement the concept design process on the basis of model-based systems engineering (MBSE), and evaluate the exchange of digital requirements and system architecture models in collaborative product development activities.

Digital development in the field of domestic aero engines has gradually received attention in recent years, especially since the establishment of the China Aero Engine Group. The full launch of the China Aero Engine Operation Management System (AEOS) and the construction of product research and development, manufacturing, supplier management, and service guarantee systems were conducted with the aims of forming a complete set of business processes and management norms that cover the entire product life cycle by 2022, forming a system architecture, improving practice methodology, strengthening collaboration, completing the digital operation of the entire system, and promoting business toward digital, networked and intelligent operation. The domestic aero engine industry in China initially formed an IPT team organization mode, a research and development plan driven by requirements, and collaborative research and development and control based on a product data center and completed the construction of document-based systems engineering. Moreover, the China Aero Engine Group has gradually promoted more advanced digital development directions, such as MBSE, digital threads, and digital twins, from the demonstration stage, and each engine research institute has established a systems engineering department to promote these directions.

Challenges and Needs

Digital engine research aims to build an advanced product digitalization platform that can be applied to the product development lifecycle, which would fundamentally change the traditional product development model, thereby significantly improving product quality, reducing costs and cycle lengths, and meeting the following requirements:

1. Building a highly reliable foundation model library and algorithm library

The new digital transformation is built on the existing mature forward research and development system, which is an inheritance and further development of the existing system and is the result of a mature design system. China needs to review the basic scientific research results and model design development experience for aero engines; establish a knowledge system and knowledge application mechanism for aero engines; develop a unified management, access, and processing knowledge system platform environment; build a comprehensive, standardized, and independent intellectual property rights advanced aero engine simulation analysis system; develop a professional model library and algorithm library that integrate basic data, calculation tools, and verification methods for engineering design requirements; build a standardized model entry system for verification, validation, and accreditation (VV&A); establish a model credibility evaluation workflow and norms; and comprehensively evaluate the effectiveness of modeling work and results at all stages.

2. Building a plug-and-play interconnected general simulation platform

All digital results must be integrated into the platform for development, operation, and expression. A plug-and-play interconnected platform is the foundation. A general simulation platform that meets the needs of aero engine professional module development and verification can be built, which can carry out module development and deployment; support multiple development languages and development frameworks; provide elastic expansion, high reliability, and a containerized program running environment; provide open data interfaces; dock with various third-party applications; become a subplatform of the product lifecycle development platform; provide visual development kits for aero engine-specific scenarios; support rapid reuse of code and different functional modules; provide open API interfaces for modules; quickly assemble and form professional modules for different application scenarios; provide data processing workflows; provide flexible support for data processing in different application scenarios; provide strong data storage and management capabilities, including for structured databases and unstructured databases, as well as time series databases, and to support file and object storage; provide a running environment based on real-time data streams; support tools and algorithms for model building on the basis of data to run on the platform; support rapid customization of data analysis results visualization, support diagnosis, evaluation, and prediction on the basis of real-time data; provide management modules, including identity management modules, permission management modules, intellectual property rights management modules, the platform's own functional modules, professional modules, and visualization modules to form standardized integration methods and norms; achieve modeling and simulation collaboration; provide unified modeling and monitoring of business processes through the use of unified business process integration tools and manage and optimize processes on this basis; and integrate tools and data

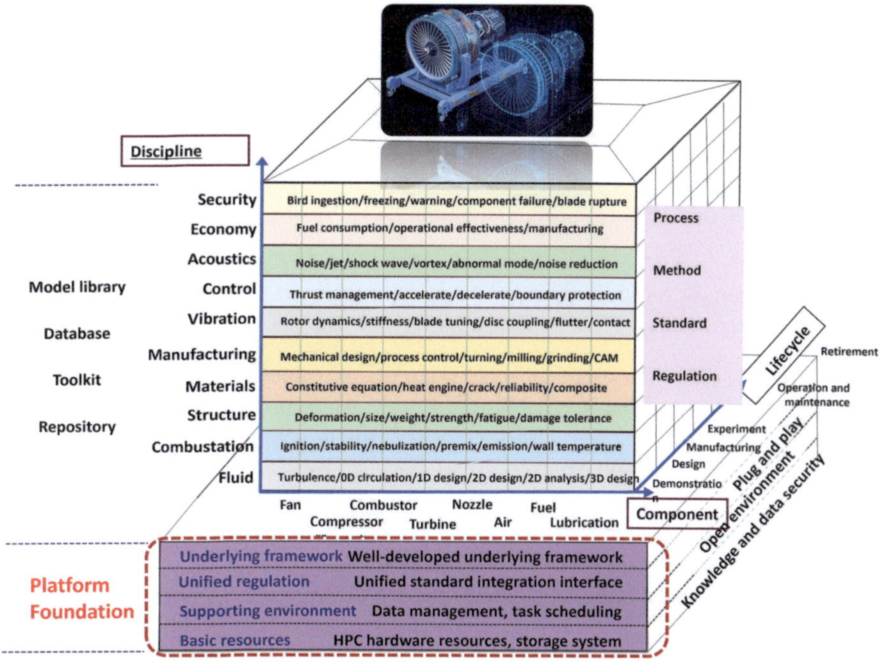

Fig. 5 Overall architecture of the digital engine simulation platform

related to digital threads into MBSE, thereby forming a semantic and connectivity-oriented innovative platform in line with the trend of digital development. The overall architecture of the digital engine platform is shown in Fig. 5, which includes the platform infrastructure, an aero engine model library, a database, a tool library, and a knowledge library (referred to as "four libraries and one platform").

3. Simulation workflow management and data processing technology based on a digital thread

A structured, formalized description system oriented to requirements and system models is formed, product information at different stages is expressed on the basis of system models, and system design and product development are continuously verified and confirmed throughout the lifecycle. Universities, research institutions, and industries work together to establish standards for aero engine requirements and system modeling; break through data exchange and interoperability standards for different tools; build system model use cases and standardized databases for different engines; connect the business chain among system modeling tools, requirement management tools, and joint simulation frameworks; realize a universal platform architecture with the system model as the only source of truth, which can be part of the digital thread that connects end-to-end data flow; define standards or norms for digital thread-related model creation, information transmission, data updates, etc.; integrate digital thread tool chains; ensure the quality of data entry and transmission throughout the lifecycle; realize effective communication of data flow; improve process data (simulation data, experimental data, etc.) management capabilities and data analysis and heterogeneous data integration capabilities; realize multisource data fusion with the system model as the source of truth; develop new process automation methods; develop interface, CAD-CAE automation, and grid automation tools; give attention to digital technologies related to intelligent recognition, behavior decision-making, and process control; and perform product technical status traceability and rapid iteration updates.

4. Development of multidisciplinary coupling integrated machine simulation technology

To meet the design requirements of aero engines, we establish a multilevel (component level, whole-machine level, etc.) and multidisciplinary integrated simulation platform that connects multidisciplinary tool chains, develop efficient algorithms for multiphysical phenomenon coupling, and develop mixed-dimensional simulation methods for whole machines and component systems. We develop efficient design space exploration and performance improvement methods and tools, which form multidisciplinary integrated optimization tools for design requirements; develop and verify closely coupled integrated verification platforms; develop high-confidence dimension reduction simulation algorithms and data-based rapid simulation methods for functions and behaviors at different abstraction levels; and develop fully digital, semiphysical and virtual–real coupled test verification systems.

5. Development of data-based analysis and decision support technologies

We build data storage and management methods and platforms, define rules and standards for processing heterogeneous aero engine data, develop tools for automated data processing and tagging and management of large-scale data, develop automated data conversion and mapping tools on the basis of rules and models, develop artificial intelligence reduction algorithms for high-dimensional simulations, develop intelligent algorithms and simulation analysis methods on the basis of data, develop modeling and simulation analysis methods that integrate mechanisms and data, develop intelligent recognition and decision support methods, and develop interactive tools for data visualization, algorithm model visualization, and supporting human decision-making and expression.

Key Technical Issues

Discipline Simulation Model Integration

For the development of the aviation industry and the pursuit of revolutionary solutions, the method of integrating models from related disciplines via physics-based analysis capabilities to obtain system-level functions and behaviors has been widely recognized by aviation practitioners. After years of development, integration technology has evolved from the single-tool integration stage to the cross-tool integration stage in a collaborative environment. Regardless of the stage, the information interaction between models from different disciplines is always a key issue in integration. The information interaction of these models includes mainly the interaction of discipline knowledge, the interaction of heterogeneous data models, and the communication of information between models. For disciplines such as aero engines with high discipline complexity and strong coupling, the efficient integration, reuse, and optimization of models is always an important research direction.

To enhance the final integration and acceptance testing capabilities of manufacturers' models, identify defects in models from different disciplines in a timely manner, reduce the probability of misunderstandings in knowledge interaction caused by semantic differences between disciplines, and enhance the effective transmission of knowledge and information between multidisciplinary models, the German Aerospace Center (DLR) began developing a technology in 2005 that enables any number of cross-disciplinary analysis capabilities to be effectively interconnected, which is called the central data model common parametric aircraft configuration scheme (CPACS) [20]. This data model is oriented toward the entire aviation industry and uses general parametric technology to provide high-fidelity descriptions of aircraft configuration schemes, as shown in Fig. 6. On the basis of the development of the CPACS and its associated tools, to promote the application of parametric aircraft configuration schemes (CPACSs) in engines, aerodynamics and other professions and disciplines; reduce the inconsistency risk of non standardized methods of exchanging configuration scheme data; and promote the integration and reuse of cross-disciplinary knowledge and models, Alder and others [21] developed the CPACSLibrary software library. The software library implements methodologies, test-driven development, version control, binding with Python, and data visualization strategies through basic programming languages such as C++, thereby accelerating the improvement of the CPACS ecosystem.

In addition to the research and development of industry-oriented general languages such as CPACS, in terms of direct knowledge and information transfer of discipline models, Sirin and others [22] proposed the model identity card (MIC) method, which forms standardized terms through the abstraction of product knowledge and discipline models, realizes the statement of different properties of models, and enhances the reliability of discipline model integration.

To simplify the creation, storage, exchange, and reuse of dynamic models of different discipline simulation systems for use in model/software/hardware in-the-loop simulation, cyber-physical systems, and other applications, initiated by Daimler AG, with the joint efforts of many industrial software manufacturers, producers, and the Modelica Association, a set of discipline interaction interface standards, named the functional mock-up interface (FMI), was formed. The FMI [23] is a tool-independent standard that supports model exchange and cosimulation of dynamic models using a combination of XML files and C code (either compiled in DLL/shared libraries or compiled in source code).

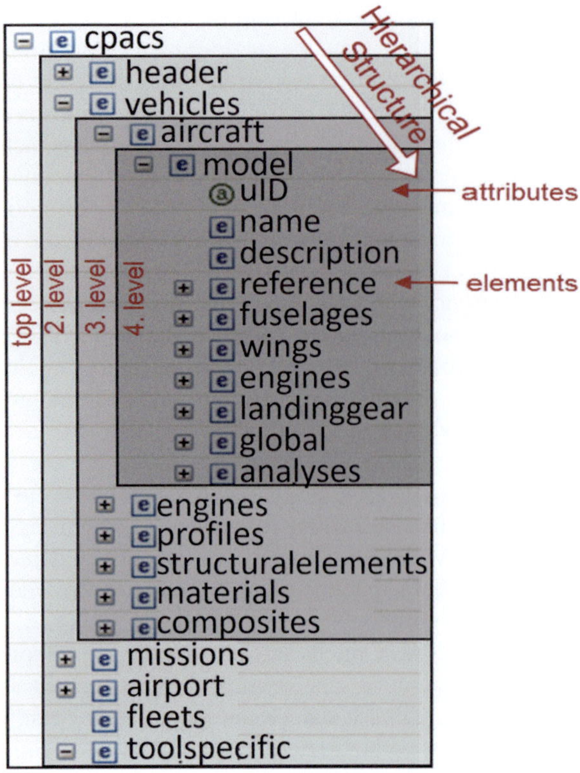

Fig. 6 CPACS structural diagram [20]

With the continuous development of integration technologies for models from various disciplines, the aviation industry, owing to the high complexity of its engineering, has made the demand for cross-engineer, cross-department, and cross-manufacturer collaborative research and development increasingly apparent and urgent. Correspondingly, the demand for the collaborative development of multidisciplinary models of different sources and different qualities also urgently needs to be satisfied. As a result, the integration of disciplinary models has gradually entered the stage of cross-tool integration in a collaborative environment. Under such conditions, the integration of disciplinary models not only faces the problems of the disciplines but also requires the definition of disciplinary workflows; the division of work responsibilities; the storage, tracing, and processing of data; and the security of the intellectual property rights of disciplinary models to be addressed. To alleviate these problems, the European Union, through the funded AGILE project, proposed the remote component environment (RCE) tool [24], which is an open-source application developed by the German Aerospace Center (DLR). Users can intuitively integrate disciplinary tools, define their dependencies through a graphical interface, and execute the resulting multidisciplinary engineering workflows. All the generated data are stored centrally for traceability, subsequent analysis, and postprocessing. To some extent, RCE enables collaborating engineers to easily contribute their disciplinary tools to multidisciplinary design or analysis, which simplifies the analysis of workflow results. In addition to RCE, many collaborative work environments across tools have been formed both domestically and abroad, such as the iDesigner platform of China Shipbuilding Industry Corporation's Orlanto and the SYSWARE platform of Suo Company. These platforms are based on the graphical integration of disciplinary models; the incorporation of industry R&D experience; and the trends of platformization, collaboration, and cross-tool and cross-regional integration of disciplinary models.

Connection of System Models and Disciplinary Models

System models can be viewed as strategic organizations of system functional elements, and their layouts enable the roles, relationships, dependencies, and interfaces between elements to be clearly defined and understood [25]. Currently, system models mainly use system modeling languages such as UML and SysML to describe the composition and association of products from a global perspective in the form of semantic modeling, but they do not contain all the detailed quantitative information on the products, which still has a certain degree of uncertainty, thus limiting the predictive ability of the system models [26]. To overcome the uncertainty problem of system models, the introduction of disciplinary models into the system modeling process, with quantitative disciplinary descriptions used to increase the effectiveness of system model decision-making and prediction, has become a widely accepted solution. The integration of disciplinary models into system models, in addition to improving the capabilities of the system models, can also provide management for disciplinary models [27], thus driving or guiding the integration and optimization of disciplinary models from the perspective of cross-disciplinary relationships and providing a new integration approach for disciplinary models for products such as aero engines with high disciplinary complexity and strong coupling.

To achieve the joint application of system models and disciplinary models, the connectivity of information between models is a key issue that must be resolved. Owing to the obvious differences between the semantic uncertainty of system models and the descriptive accuracy of disciplinary models, in addition to clarifying interaction interface specifications, data types, and communication rules, it is also necessary to carefully design and define the attributes of information interaction and storage in each model.

Currently, for research on the connection between system models and disciplinary simulation models, many studies have borrowed the idea of integration between disciplinary models, such as that of Galisson and others [28], who used the Arcadia methodology in the process of creating product system models with the Capella tool, applied the model identification card (MIC) method, and introduced a new viewpoint configuration into the system model to capture the requirements and features of disciplinary simulation, thus ultimately forming an MIC oriented toward disciplinary modeling and evaluation and achieving interaction between system models and disciplinary models.

For disciplinary geometric models that are still centered on document management, many teams have attempted to use system models to drive and manage geometric models to enhance the ability to query and analyze system models and disciplinary geometric models. Bajaj and others [29], in the system modeling process, associated geometric models with requirements in a parametric modeling manner. They required the "seed" to become the starting point of the disciplinary geometric model, encoded system-level constraints as geometric or parameter boundaries, visualized system requirements as geometric figures, and realized a method for dynamically calculating whether the geometric model violates constraints. In addition, by extracting system attributes from the baseline model of the disciplinary geometric model at different stages, automated verification of engineering analysis and system requirements was achieved. After a series of studies and applications in the direction of digital threads, Bajaj and others [30] developed a system lifecycle handling (SLH) software environment in the Syndeia platform, which was launched by Intercax in the United States, to provide services for building, managing, querying, and visualizing digital threads by connecting heterogeneous artifacts from requirements and the system architecture to PLM/computer-aided design, manufacturing and simulation models to machine and sensor data streams. Syndeia uses open standards such as STEP, REST/HTTP, and OSLC to combine models and data from various software repositories and tools, such as SysML tools, PLM and ALM environments, databases, requirements and project management systems, and simulation tools, and provides visualization services for compiling overall system model diagrams of products/systems for each repository, tool, model, and data item to ensure information association, traceability, and smooth interaction while minimizing the learning and usage costs of engineers.

Thomas [2] introduced the concept of a multidisciplinary digital integration chain through in-depth research on the geometric model of the discipline. The multidisciplinary digital integration chain is a multilevel design–simulation loop that integrates subsystem models and data (potentially from multiple disciplines) to predict global system behavior and verify whether system performance meets expectations. In the context of the digital integration chain, a digital prototype conversion method suitable for large assembly finite element models is proposed, which makes the system model applicable to the behavioral simulation of the discipline geometric model.

On the basis of existing research, the European-funded AGILE4.0 project also focuses on the integration of system models and discipline models, with the aim of accelerating the efficiency of mutual transformation between system models and discipline models while achieving system scheme trade-offs and requirement traceability [31]. Through research and practice, a development framework that combines MBSE and MDAO has been formed [32]. Under this framework, a series of system model development and verification tools, discipline model development and analysis tools, and data specifications for supporting the transformation of system models and discipline models have been integrated [33], and applications have been carried out in areas such as drone certification [34] and large aircraft supply chain management [35].

Simulation Modeling Method for a Digital Twin

Multiple stages in the product lifecycle provide information to the digital thread, which can be used to make wise choices in future designs and reduce the uncertainty of design parameters and process costs. In addition, this information may reveal more effective operating strategies. Making design decisions adds new information to the product lifecycle, thus changing the state of the digital thread. This process can be described mathematically using data-driven design methods and decision problems under uncertainty. Tools such as Bayesian inference and decision theory can be used for mathematical description [36].

For the digital twin modeling problem of aero engines, early methods include mainly mechanistic modeling, data-driven modeling, and mixed mechanistic/data modeling. Mechanistic modeling starts from the mechanism of the aero engine, which is based on general theory models, and applies thermodynamic formulas of the aero engine cycle to model the intake, compressor, combustion chamber, turbine, and tailpipe, with consideration of the flow equation in the thermodynamic cycle, the compression equation in the compressor, the heat balance equation in the combustion chamber, the heat exchange equation, etc., to establish the digital twin model of the aero engine. The shortcoming of mechanistic modeling is the lack of integration with the actual data of the aero engine, which makes accurate description and prediction of the performance characteristics of the aero engine throughout its lifecycle difficult.

In recent years, with the development of data-driven modeling theory, including support vector regression modeling, extreme learning machine modeling, intelligent identification optimization algorithms, and basic theories of machine learning, scholars outside of China have also developed many high-precision digital twin models directly identified/trained on equipment operation data, such as the NARMAX identification method, which can perform parameter analysis on the aero engine in the time domain and frequency domain, determine the model order, and model the aero engine, thereby further improving the accuracy of the modeling method. The mainstream data-driven modeling methods are neural network methods, and their powerful mapping ability can realize multivariable, strongly coupled, nonlinear system modeling. Data-driven modeling selects a large number of sample data and uses neural network algorithms for training, which can achieve better simulation results, and the calculation results usually accurately match the actual data. The shortcoming of data-driven modeling is that it ignores the physical characteristics of the equipment; thus, it is difficult to provide a mechanistic explanation of the results.

For the digital twin simulation modeling of an aero engine, when the use scenario requires a high degree of confidence, a single modeling method obviously cannot accurately reflect the overall performance of the aero engine or even the entire aircraft in the full envelope or the full state range, and the method of data/mechanism mixed modeling should be used. However, owing to the high fidelity requirements for the entire digital twin, industrial applications require not only thermodynamic performance information for the engine but also information on its structure, internal three-dimensional flow, and even performance under the entire lifecycle or various faults. Traditional single-discipline simulation cannot meet the application requirements of digital twins for aero engines, and matching between multiple professions has become a standard configuration for digital twins; thus, the modeling methods tend to be multidisciplinary and cross-professional and need to be implemented through model order reduction technology, general interface model design, and unified simulation architecture design. The aero engine aerodynamics, structure, system and other professional models and finite element simulation models are effectively integrated to achieve a system-level multidisciplinary joint simulation model that involves various modeling technologies, including modeling specification construction, multidisciplinary system modeling, finite element model reduction, heterogeneous model integration, model correction and update, virtual–real interaction development, and model real-time technology.

Conclusions

Owing to the difficulty and complexity of aero engine products, support from digital technology is urgently required. For aviation powers in Europe and America, digital engine simulation technology has played a key role in the development of aero engines, work on data and model connectivity is steadily advancing, and high-precision digital modeling and dynamic simulation technology are running through the entire lifecycle of aero engine R&D and maintenance and have been gradually integrated into forward design R&D systems, which has led to significant efficiency improvements and cost reductions in engine development. Promoting the transformation of engine equipment from long-term trials to rapid iterations, building low-cost sustainable supply capabilities, and constructing a digital ecosystem for aviation engines are tasks of urgent necessity.

By reviewing the development history and substantive content of digitalization concepts, this article highlights the close and inseparable relationship among MBSE, digital twins, digital threads and other technologies and discusses how the problems faced by digitalization are not unique to aviation engines; however, the solutions must be dominated by industry characteristics and expert knowledge combined with the concept of digital engineering to develop digital engines. Through the analysis of the development status of the industry in China and other countries, in the face of the challenges with digital engines, the work that needs to be carried out is summarized and includes the development of a highly reliable basic model library and algorithm library, a plug-and-play interconnected general simulation platform, simulation workflow management and data processing technology based on the digital mainline, multidisciplinary coupling whole machine integration simulation technology, data-based analysis and decision support technology. For the key technical issues that need to be addressed for digitalization, this paper summarizes the research content and directions of discipline simulation integration methods, system model and discipline model connection methods, and digital twin modeling methods in China and other countries to provide support for the industry to carry out specific technical development work.

To carry out research on digital engines and build a digital collaborative platform for the entire life cycle of aviation engines, the integration of the latest achievements in aviation engine design technology, simulation technology and information technology is necessary, along with the wide and deep cooperation of universities, research institutes, manufacturers and other units with disruptive digitalization, intelligent concepts and specialized aviation engine professional knowledge.

References

1. Cao JG (2018) Status, Challenges and perspectives of aero-engine simulation technology. Propuls Technol 39(5):961–970
2. Vosgien T, Van TN, Jankovic M et al (2012) Towards model-based system engineering for simulation-based design in product data management systems. IFIP Adv Inf Commun Technol 388(2):612–622
3. Herzig SJI, Qamar A, Reichwein A et al (2011) A conceptual framework for consistency management in model-based systems engineering. In: 31st Computers and information in engineering conference, Parts A and B, vol 2. Washington, DC, USA: ASMEDC, pp 1329–1339
4. Roper W (2020) Take the red pill: the new digital acquisition reality. Department of the Air Force
5. Feng G, Ping G, Guodong Z, Yan K, Junjie R, Degang J (2022) Development trend of MBSE and investigation of concurrent collaborative demonstration for Chinese lunar exploration program. Chin J Space Sci 42(02):183–190
6. Guan F, Ge P, Zhou GD et al (2022) Development trend of MBSE and investigation of concurrent collaborative demonstration for Chinese lunar exploration program. Chin J Space Sci 42(02):183–190
7. Che Y (2020) Application of MBSE in aero-engine structural parameter calculating system. In: The 32 nd congress of the international council of the aeronautical sciences. Shanghai, China: ICAS, pp 3–10
8. Ting L, Jianchao Z (2021) Preliminary discussion on application of digital thread to aero engine. Aerosp Power Technol 02:30–34
9. Liu T, Zhang JC (2021) Preliminary discussion on application of digital thread to aero engine. Aerosp Power Technol 02:30–34
10. Lu J, Chen DJ, Gürdür D et al (2017) An investigation of functionalities of future tool-chain for aerospace industry. INCOSE Int Symp 27(1):1408–1422
11. Tao F, Zhang H, Liu A et al (2019) Digital twin in industry: state-of-the-art. IEEE Trans Industr Inf 15(4):2405–2415
12. Xie R, Chen M, Liu W et al (2021) Digital twin technologies for turbomachinery in a life cycle perspective: a review. Sustainability 13(5):2495
13. Zhuyin R, Hua Z, Jian Z, Qi Z (2022) Application and prospect of digital twin in design phase of aero-engine combustion chambers. Aeronaut Manuf Technol 65(17):34–39
14. Lo CK, Chen CH, Zhong RY (2021) A review of digital twin in product design and development. Adv Eng Inform 48:101297

15. Zaccaria V, Stenfelt M, Aslanidou I et al (2018) Fleet monitoring and diagnostics framework based on digital twin of aero-engines. In: Turbo expo: power for land, sea, and air, vol 51128. American Society of Mechanical Engineers. https://doi.org/10.1115/GT2018-76414
16. Ferrari A, Nagel A (2020) Connecting, tracing and managing the lifecycle of models, simulation and linked data: is that easy? Raytheon Technologies
17. Ren ZY, Zhou H, Zhang J et al (2022) Application and prospect of digital twin in design phase of aero-engine combustion chambers. Aeronaut Manuf Technol 65(17):34–39
18. Noguchi RA, Wheaton MJ, Martin JN (2020) Digital engineering strategy to enable enterprise systems engineering. INCOSE Int Symp 30(1):1727–1741
19. Moitra A (2022) Validation and verification for digital twins. GE Research
20. Zolotas A, Hoyos Rodriguez H, Hutchesson S et al (2020) Bridging proprietary modelling and open-source model management tools: the case of PTC integrity modeller and epsilon. Softw Syst Model 19(1):17–38
21. Lacaille J (2021) Advances in condition monitoring and structural health monitoring. Singapore: Springer Singapore
22. Salnikov AV, Gordin MV, Shmotin Yu N et al (2022) Digital twins—a platform for aircraft engine lifecycle management. In: Proceedings of higher educational institutions machine building. https://doi.org/10.18698/0536-1044-2022-4-60-72
23. Nagel B, Böhnke D, Gollnick V et al (2012) Communication in aircraft design: can we establish a common language. In: 28th international congress of the aeronautical sciences, Brisbane, Australia: ICAS, pp 443–455
24. Alder M, Skopnik A (2022) Development of a software library for performant and consistent CPACS data processing. In: Deutscher luft- und raumfahrtkongress, Dresden: DGLR
25. Sirin G, Paredis CJJ, Yannou B et al (2015) A model identity card to support simulation model development process in a collaborative multidisciplinary design environment. IEEE Syst J 9(4):1151–1162
26. Blochwitz T, Otter M, Åkesson J et al (2012) Functional mockup interface 2.0: the standard for tool independent exchange of simulation models. In: Proceedings of the 9th international MODELICA conference. Munich, Germany, pp 173–184. https://doi.org/10.3384/ecp12076173
27. Boden B, Flink J, Först N et al (2021) RCE: an integration environment for engineering and science. SoftwareX 15:100759
28. Kapurch SJ (2017) NASA systems engineering handbook. Diane Publishing, USA
29. Nowodzienski P, Navas J (2022) From model-based to model and simulation-based systems architectures—achieving quality engineering through descriptive and analytical models. INCOSE Int Symp 32(1):1247–1266
30. Watson MD, Mesmer BL, Farrington PA (2020) Engineering elegant systems: theory of systems engineering. Huntsville, Alabama, USA: NASA
31. Galisson G, Hammadi M, Gherib M et al (2022) Model-based systems engineering methodology for defining multi-physics simulation models. In: 2022 IEEE international systems conference (SysCon). Montreal, QC, Canada: IEEE, pp 1–8
32. Bajaj M, Cole B, Zwemer D (2016) Architecture to geometry—integrating system models with mechanical design. In: AIAA SPACE 2016. Long Beach, California: American Institute of Aeronautics and Astronautics, 2016. https://doi.org/10.2514/6.2016-5470
33. Bajaj PM, Hedberg T Jr (2018) System lifecycle handler—spinning a digital thread for manufacturing. INCOSE Int Symp 28(1):1636–1650
34. Ciampa PD, Nagel B (2021) Accelerating the development of complex systems in aeronautics via MBSE and MDAO: a roadmap to agility. In: AIAA aviation 2021 forum, VIRTUAL EVENT: American Institute of Aeronautics and Astronautics. https://doi.org/10.2514/6.2021-3056
35. Boggero L, Lefèbvre T, Vankan J et al (2022) The agile 4.0 MBSE-MDAO development framework: overview and assessment. In: 33rd ICAS congress, Stockholm, Sweden: ICAS
36. Ciampa PD, La Rocca G, Nagel B (2020) A MBSE approach to MDAO systems for the development of complex products. In: AIAA aviation 2020 forum. virtual event: American Institute of Aeronautics and Astronautics. https://doi.org/10.2514/6.2020-3150

Simulation Modeling Method for Aero Engines Based on Digital Twins

Abstract This chapter reviews digital twin-based simulation modeling methods for aero engines, which integrate physical model-driven and data-driven approaches to achieve high-fidelity simulations of aero engine behavior mechanisms. This method first establishes a performance mechanism model of the target engine and uses bench test and flight data to adaptively refine it using machine learning algorithms to obtain a digital twin model for simulation. This model can reflect the physical characteristics and dynamic behavior of the engine, thereby enabling precise monitoring and prediction of engine performance. In addition, this method combines artificial intelligence algorithms to analyze the modeling bias compensation mechanism, thereby improving the reliability of aero engine simulation and providing strong support for engine performance optimization, performance prediction, and fault diagnosis.

Keywords Aero engine · Digital twin · Artificial intelligence · Simulation modeling · Bias compensation

Introduction

Significance

As the "heart" of modern aircraft, both the performance and reliability of aero engines govern the safety, economy, and environmental protection of flight. Traditional modeling methods rely mainly on physical experiments and simplified mathematical models. When dynamic behavior analysis is performed under complex working conditions, such as extreme temperature environments, coupled vibration phenomena, and material fatigue problems, these methods often fall short. Simulation modeling technology, which is based on digital twins, can deeply integrate a physical model of an engine with actual operating data to reflect the operating status of the engine in real time through a high- fidelity virtual model.

By building a real-time bidirectional synchronization system between the physical and virtual domains, digital twin technology can not only accurately simulate the

Q. Xu et al., *Digital Intelligent Aeroengine*,
https://doi.org/10.1007/978-981-95-0942-3_2

performance of the engine under across operational envelopes but also via optimization algorithms for real-time control parameter adaptation, thereby achieving closed-loop optimization of engine performance. The digital twin model, which uses historical data and machine learning algorithms, can accurately predict the performance degradation trend and service life of an engine. This predictive capability facilitates proactive maintenance strategy development and extend the service life through operational parameter optimization, thereby holistically enhancing overall performance via closed-loop control. In addition, during the design phase, digital twin technology can quickly verify the performance of various design schemes through virtual simulation and multidisciplinary optimization methods, thus effectively shortening the research and development (R&D) cycle and reducing R&D costs.

Digital twin technology provides disruptive tools for the full lifecycle management of aero engines (covering design, manufacturing, and maintenance). The various functions and compositions of the digital twin aero engine [1] are detailed in Fig. 1. By conducting multiple-condition verification and virtual experiments, potential problems can be discovered before the engine is put into use and actual test plans can be optimized, thereby further improving the performance and reliability level of the engine. In summary, simulation modeling based on digital twin technology provides a comprehensive and efficient technical approach for optimizing aero engine performance, which strongly promotes the digital transformation process of aero engine technology.

Development History of Digital Twin Technology

Digital twin technology constructs digital models that mirror real-world physical objects or processes in virtual space to achieve accurate mapping and efficient management of the physical world. This technology comprehensively uses sensors, the Internet of Things, big data, artificial intelligence, and virtual reality to describe and model the characteristics, behavior, and operation process of physical objects and accurately reflects the status and behavior of physical entities through real-time data updates.

The core of the digital twin consists of three parts: physical entities, digital models, and data connections. Physical entities refer to objects or systems in the real world; digital models are their virtual replicas, which are used for simulation and analysis; and data connections realize the bidirectional interaction between physical and digital spaces through sensors and Internet of Things technology. Digital twins are data-driven and provide model support, software definition, and precise mapping, and they can achieve seamless connections between the physical world and the digital world.

The concept of the digital twin originated in the 1960s. At that time, NASA used multiple simulators to simulate the operating status of spacecraft in the Apollo program to assist in astronaut training and ground control personnel decision-making.

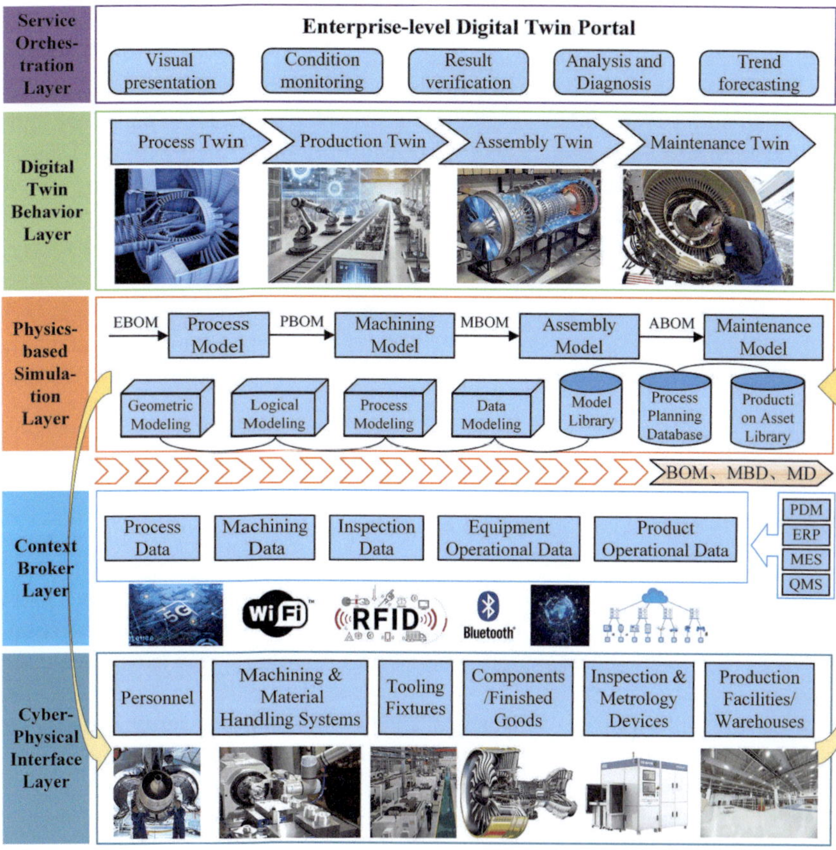

Fig. 1 Composition of the digital twin aero engine [1]

These simulators maintained a close connection with the spacecraft through real-time data transmission and laid the foundation for the birth of digital twins.

In 2002, Professor Michael Grieves of the University of Michigan first proposed the concept of "virtual digital mapping of physical products", which was a prototype of digital twins. Although this concept was not widely recognized at the time, it provided theoretical support for subsequent development. With the rapid development of technologies such as the Internet of Things, big data, and artificial intelligence, digital twin technology has reached technical maturity. In 2003, Professor Grieves further improved the concept of the digital twin by proposing the use of digital models for simulation analysis and testing of physical entities.

The 2010s were a crucial period for digital twin technology, in which it transitioned from concept to preliminary applications. In 2010, NASA first applied the concept of digital twins to the aerospace industry, aiming to enable real-time monitoring and simulation of spacecraft performance by constructing a digital map of the spacecraft and optimizing the reliability and overall efficiency of the system. During this period, digital twin technology was applied mainly in high-tech fields such as aerospace and defense, with the core goal being to achieve virtual mapping of physical entities through model simulation-driven and perception control-driven methods.

At the beginning of the 2020s, digital twin technology entered a period of rapid development. At this stage, the technology was deeply integrated with the Internet of Things, big data, artificial intelligence, and cloud computing, thus forming a more powerful technical system architecture. IoT technology enables interconnection and interoperability between physical devices, big data technology supports the storage and analysis of massive amounts of data, and artificial intelligence technology focuses on model optimization and prediction. The application scenarios of digital twin technology continue to expand, and the market size is growing rapidly. Digital twin technology is becoming a key force that drives the digital transformation of various industries.

Development History of Aircraft Engine Simulation Modeling

The early exploration of aircraft engine simulation modeling began in the 1980s, with research focused on the development and initial application of numerical simulation technology. During this period, simulation technology was used mainly for evaluating and optimizing design schemes during the engine design stage to provide support for design through one-dimensional and two-dimensional aerothermal simulations. However, owing to computational limitations, the accuracy of early simulation models was low, and they were mostly single-discipline models, which made meeting the multidisciplinary coupling needs of complex systems difficult.

In the 2010s, with the advancement of high-performance computing technology, aircraft engine simulation modeling entered a systematic development stage. The Numerical Propulsion System Simulation (NPSS) program in the United States is a typical representative of this period. The program was led by NASA and jointly promoted by industry, universities, and government agencies, and many resources were invested into the research and development of simulation technology. During this period, simulation technology gradually expanded from one-dimensional and two-dimensional methods to three-dimensional methods, thus achieving the multidisciplinary coupling simulation of entire machines and making significant progress in the integrated simulation of key components such as fans, compressors, and combustion chambers.

From the early 2020s to the present, aircraft engine simulation modeling technology has been deeply integrated with emerging technologies such as big data, artificial intelligence, and the Internet of Things, and enabled a shift from virtual testing to full lifecycle management. Simulation technology is not only applied in the design phase but also spans the entire lifecycle of an engine, including engineering development, performance optimization, and fault diagnosis. For example, when combined with physical test data, a simulation model can achieve more accurate performance prediction and optimization, significantly reducing physical testing costs and efforts.

In addition, simulation technology is also used to investigate rare faults and operating conditions that are difficult to replicate, which makes it an indispensable tool in aircraft engine research and development.

Aircraft Engine Digital Twin Modeling Method

The digital twin-based aircraft engine simulation modeling approach, integrating physical model-driven methodologies with data-driven techniques, enables high-fidelity simulations of engine performance mechanisms. This method first establishes a performance mechanism model of the target engine and uses bench test and flight data through machine learning algorithms for adaptive correction to obtain a digital twin model for simulation. This model can reflect the physical characteristics and dynamic behavior of the engine and support accurate detection and prediction of engine performance. In addition, in this method, artificial intelligence algorithms are combined to analyze the modeling bias compensation mechanism, thereby improving the reliability of aircraft engine simulation and providing strong support for engine performance optimization, performance prediction, and fault diagnosis.

Construction of the Performance Mechanism Model

The modeling of the multidisciplinary coupled performance mechanism model of an aircraft engine involves the comprehensive consideration of multiple disciplines, such as fluid mechanics, thermodynamics, structural mechanics, vibration theory, and control theory. The mutual influence among these disciplines makes the performance of the aircraft engine complex and difficult to predict; therefore, accurately describing its working mechanism through a multidisciplinary coupling model is necessary.

During the modeling process, fluid mechanics models characterize gas flow behavior within engine components (such as compressors, combustion chambers, and turbines), which involves the distribution of parameters such as gas velocity, temperature, and pressure. Through computational fluid dynamics (CFD) methods, the flow behavior of the gas inside each component can be simulated, and the combustion effects on gas thermodynamic states can be further analyzed in conjunction with a combustion model. A thermodynamic model is used to calculate the inter-component gas energy transformation, study the changes in thermodynamic parameters such as temperature, pressure, and density, and thus predict the overall thermal efficiency and performance of the engine.

A structural mechanics model is employed to characterize the mechanical responses of various components of the engine under working conditions, especially for the analysis of the stress, deformation, and fatigue life of key components such as compressor blades and turbine blades. Since these components need to withstand high temperatures, high pressures, and high-speed airflow, accurate modeling of their material properties and structural design is crucial.

A vibration model is used to predict the potential vibration phenomena during engine operation, especially in high-speed rotating compressors and turbine components, where vibration problems may lead to structural fatigue or even failure. Therefore, vibration analysis can help designers identify potential vibration problems in advance and mitigate the impact of vibration on engine performance through optimized design.

The control model focuses on regulating the engine system during engine operation, especially the real-time adjustment of parameters such as combustion, airflow, and speed. The control system model needs to consider not only the working characteristics of the engine but also the effects of external environmental changes (e.g., ambient temperature, altitude, and flight mode) on engine performance to ensure that the engine can maintain optimal operating conditions under different working conditions.

Finally, for the establishment of a multidisciplinary coupled performance mechanism model, the physical behavior of each subsystem must be considered, and the subsystem behaviors must be coupled using suitable mathematical techniques to form a comprehensive overall model. Common coupling methods include system-level simulation, numerical optimization, and multidisciplinary design optimization (MDO) techniques. Through the establishment of a multidisciplinary coupled performance mechanism model, it is possible to predict the performance of the aero engine comprehensively and accurately to provide a theoretical basis and support the design, optimization, and fault diagnosis of the engine.

Within the specialized field of constructing aero-engine behavioral mechanism models, extensive and in-depth research has been conducted by scholars globally. Song et al. [2] proposed a multicell LPV model based on an aero-engine Linear Parameter-Varying (LPV) framework. Their approach involved: first, establishing an engine linear model using Jacobian linearization; subsequently deriving an affine parameter-dependent model through fitting. Then, transforming it into a multicell model via multicell theory, with flight altitude and Mach number selected as scheduling variables. The multicell LPV model was benchmarked against the nonlinear model in simulations. Results demonstrated that the established quasi-LPV model exhibits superior dynamic and tracking performance compared to the nonlinear model, thereby offering a solid foundation for research into LPV gain control of aero engines. Zheng et al. [3] introduced a scheme for an engine overall performance simulation model based on fluid network topology. Building upon an object-oriented component performance modeling approach and a general simulation system framework, their method leveraged iterative variables, a system of balance equations, and component calculation sequence to construct a steady-state performance simulation model applicable to diverse aero-engines and gas turbines. Simulation results demonstrated that this scheme and model enable the automated construction of the engine computational topology, iterative variables, and balance equations, thereby enhancing the applicability of the simulation system. Yang et al. [4] developed an aero engine component-level identification model based on system identification methodologies. This approach models each component as a Hammerstein system for predicting key parameters. The parameters of the dynamic linear part and the static nonlinear part within the Hammerstein system effectively capture the dynamic characteristics of the aero engine. Simulations of the engine acceleration process from ground idle to maximum thrust were conducted, with predicted data were compared against actual operational data. Results indicate that the root mean square error (RMSE) for the identified key engine parameters using this model was below 1%. Gao et al. [5] conducted performance modeling research on a certain type of dual-rotor mixed-flow turbofan engine utilizing a general object-oriented simulation framework. They developed an aero engine model accounting for factors such as non-uniform fan outlet flow distribution, high/low-pressure turbine cooling bleed air, and variable specific heat effects. Results demonstrate that the deviation between the model's calculated overall performance across various operating conditions and the measured values was below 3%. Chen et al. [6] proposed a variable baseline modeling approach based on state awareness. Building upon wavelet transform filtering, they introduced an optimal threshold calculation method for steady-state data screening to reduce data mis-selection or omission. Additionally, we introduce a variable baseline modeling model utilizing the Gaussian mixture model (GMM). Compared to conventional single baseline models, the proposed variable baseline model reduces the relative mean error for high- pressure and low-pressure rotor speeds by over 45% and 30%, respectively. Gao [7] developed a steady-state model for a dual-rotor mixed-exhaust turbofan engine and its gas turbine starter, incorporating the effects of a humid and smoky environment based on the aero-engine's operating principles. Results indicate that increasing humidity shifts the compressor characteristic line leftward. Smoke exposure exhibits analogous performance degradation, and the inlet smoke temperature must be considered.

Pang et al. [8] introduced an online state space modeling (SSM) method for aero-engines based on component-level models. This approach constructs high-fidelity dynamic models leveraging partial derivatives iteration and Taylor series expansion techniques. Innovatively employing a non-perturbative partial derivative iterative algorithm, it enables accurate computation of complex systems derivatives without traditional perturbation analysis. Furthermore, it establishes a unified SSM via Taylor expansion, capable of concurrently characterizing the dynamic behavior at both engine steady-state operating points and during transient transition processes. Simulation results demonstrate that when the input step amplitude is within 2%, the SSM retains its accuracy; when the input step amplitude is large, the SSM response closely matches that of the component-level model. Roman [9] proposed a time-domain identification method for aircraft turbofan engine models. This method utilizes a multivariate linear model representing turbofan engine dynamics with uncertain eigenvalues. Uncertainty is employed to quantify the discrepancy between nonlinear and linear models. The real and imaginary parts of the uncertain matrix eigenvalues are determined using nonlinear programming techniques (Fig. 2).

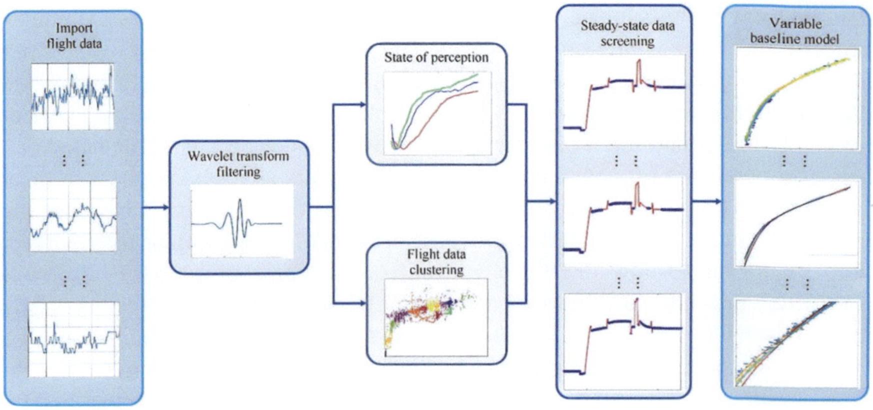

Fig. 2 Overall structure of the baseline model [6]

Data-Driven Adaptive Model Correction

Under the classical modeling paradigm, the analysis of the operating characteristics of aero power devices usually relies on a multidisciplinary coupling characterization system that is built on first principles and covers complex physical field interactions such as viscous fluid motion, thermodynamic conversion processes, and elastic body dynamics. These first-principles models govern nonlinear partial differential equations (PDEs), use numerical techniques such as the finite volume method (FVM) and isogeometric analysis (IGA) to solve for field variables, and characterize the energy conversion efficiency of core components such as compressors and combustion chambers with high fidelity under steady-state conditions. However, this modeling method is limited by idealized assumptions, and its solution accuracy decreases exponentially with increasing Reynolds number, which makes capturing nonlinear dynamics effectively under the coupling of unsteady flow and structural flutter difficult. However, because many factors in the actual operating environment (such as airflow fluctuations, temperature changes, and component wear) may cause model discrepancies, physical models may not fully reflect real system behavior under some complex conditions, which poses a serious challenge to traditional modeling methods.

The data-driven adaptive model correction of aero engines is a hybrid method that integrates traditional physical modeling and modern data-driven technology. Its goal is to improve the accuracy and robustness of aero engine performance prediction, especially the adaptability under complex conditions. Through the analysis of a large amount of actual operating data, this method can reveal the potential rules of the system, complement the existing physical model, and update model parameters dynamically or make structural corrections to address various uncertainties and changes that may occur in actual operations. This method not only optimizes the model predictive accuracy but also enhances the generalization ability of the model when facing unknown data.

Adaptive model correction is the key link in combining physical models with data-driven methods. In this process, a physical model provides the basic framework and theoretical support for the system, whereas data-driven methods are used to correct and optimize the model in real time. Usually, this adaptive correction is carried out through the following steps: first, the physical model predicts the performance of the engine on the basis of preliminary design or historical data; then, the actual data obtained through real-time monitoring (such as sensor data) are compared with the results predicted by the physical model; and finally, adaptive algorithms (e.g., Kalman filtering and Bayesian optimization) continuously adjust the model parameters so that the model can adaptively correct and gradually approach the behavior of the real system.

The global research community, particularly in China, has made significant advances in data-driven adaptive model calibration for aero-engines, with extensive studies focused on real-time performance compensation and uncertainty quantification. Li et al. [10] developed an engine performance model correction method based on Quick Access Recorder (QAR) data. Employing design-point performance matching via cyclic iterative calculations and off-design performance matching through characteristic map scaling and parameter optimization, they constructed a CFM756-7B engine fleet digital twin model using GasTurb software. This model supports aero-engine performance evaluation, fault prediction and health management. Zheng [11] leveraged a Gaussian process regression (GPR) data-driven model to establish, using the Kalman filter method, a fused parallel integration of a physics-based model and the data-driven model. Verification demonstrated that under engine component failure conditions, the fused model's parameter estimation error remains below 5%, thereby meeting the digital twin's requirements for measurable parameter estimation while satisfying its high-fidelity characteristics. Liang [12] introduced a hierarchical dynamic modeling approach for entire aero-engines based on "super model" virtual test data. Specifically, the method validates simplified single component models using supermodel virtual data, corrects component coupling relationships via interface supermodel data, and integrates the verified rotor and casing system models to form the full-engine dynamic model. For a specific turbofan engine, the critical speed prediction from the simplified full-engine model obtained via this method was compared against the reference critical speed from actual full-engine testing, yielding an error within 10%. Liu [13] employed a generalized regression neural network (GRNN) and fuzzy least squares support vector machine (FLSSVM) for identification and modeling of ground microturbine engine speed. Comparison of test data R2 and Mape values reveals that the FLSSVM achieved an R2 closer to 1 and a lower Mape than the GRNN, indicating that the FLSSVM-based microturbine engine speed prediction model exhibits superior generalization capability. Lai et al. [14] proposed an engine performance parameter prediction method based on fuzzy inference and the XGBoost algorithm. First, performing an overall engine performance analysis, they identified throttle position, pressure altitude, total temperature, gross weight, Mach number, and flight phase as the primary factors influencing engine performance parameters. Then, utilizing fuzzy inference to process quick access recorder (QAR) data, they finally optimized the model parameters using the XGBoost algorithm. This method achieved favorable results in predicting civil aviation aircraft engine performance parameters.

Xiao et al. [15] developed a novel nonlinear model predictive control (NMPC) method based on a hybrid gray wolf optimization (GWO) of a Resilient Backpropagation (BP) neural network. A resilient BP neural network trains the aircraft engine prediction model, while the GWO enhances the initial parameter selection for the resilient BP neural network to improve modeling accuracy. By incorporating a logistic chaotic sequence, an individual optimal search mechanism, and crossover operation into the standard GWO, a new Hybrid GWO (HGWO) algorithm was proposed and applied to aircraft engine NMPC. Simulation results demonstrate that the proposed NMPC delivers excellent control performance for the aircraft engine. Guo et al. [16] introduced a real-time adaptive model for key aircraft engine parameters based on an Online Sequential Extreme Learning Machine-Extended Kalman Filter (OSELM-EKF) approach. This model integrates an adaptive strategy combining OSELM and EKF. A neural network was introduced into compressor component calculations, the EKF estimates health parameters to adjust system equation residuals, and the learning machine compensates for rotating component pressure ratios under varying degradation levels. The computational accuracy and speed of the adaptive parameter model were systematically evaluated across diverse scenarios. Results indicate that the proposed method outperforms traditional adaptive correction schemes. Kim et al. [17] proposed a simulation method for predicting aircraft engine transient performance using data-driven models. Input parameters for training the models include fuel mass flow rate, altitude, ultralight Mach number, and the power required generated by rotating components inertia. Output parameters comprise engine net thrust, low-pressure shaft speed, pressure and temperature at each station, propulsion efficiency, energy conversion efficiency, and overall efficiency. Simulation results revealed R^2 values exceeding 0.98 for most outputs, whereas energy conversion efficiency and overall efficiency achieved R^2 values of 0.9715–0.9183, respectively. Pan et al. [18] presented a data-driven Takagi–Sugeno (T-S) fuzzy modeling method for the aircraft engine flight envelope. Engine dynamics clustering first determines a coarse T-S fuzzy model to approximate the engine's nonlinear dynamics within the envelope. Subsequently, the MMD-FCM algorithm refines the fuzzy rules, and the least squares method determines initial parameters of each coarse model. The proposed MMD-FCM algorithm ensures stability and rationality in the refinement results, thereby enhancing identification accuracy for both steady-state and transient phases. Validation results show that the T-S fuzzy model achieves high accuracy, steady-state error below 5%, transient root mean square error (RMSE) unde 8×10^{-4}, and exhibits strong generalization ability within the secondary envelope.

Modeling Bias Compensation Mechanism

Although data-driven corrections can significantly improve the accuracy of the model, modeling bias may still exist because of model assumptions and measurement noise. Therefore, studying the modeling bias compensation mechanism is an important means for improving simulation reliability. This is usually achieved through the development of an error model to quantify and compensate for model bias through error modeling on the basis of historical data. The error model can be trained via statistical methods (the least squares method, regression analysis, etc.) or machine learning techniques (neural networks, support vector machines, etc.) to identify the pattern of model bias and apply a correction algorithm for each prediction cycle. This method can continuously update the error model through online learning, thereby achieving real-time compensation.

The bias compensation mechanism can also be combined with multidisciplinary coupling optimization methods. The coupling effect between different disciplines may lead to the cumulative effect of individual disciplinary biases. Multidisciplinary design optimization (MDO) methods can achieve information transmission and correction between different disciplines, thereby improving the accuracy of the overall model. The bias compensation mechanism usually uses a multilevel feedback system to dynamically adjust the accuracy of the model. For example, static bias compensation is carried out during the design stage, and dynamic compensation is implemented during the production and flight test stages. Through the staged feedback mechanism, the model's predictive capability is continuously refined.

In response to the bias in the aircraft engine model, many scholars in China and other countries have widely researched bias compensation mechanisms. Li et al. [19] investigated the control plan for a certain type of aircraft engine, incorporating flight parameter data to establish a neural network-based steady-state parameter relationship model. This framework enabled simulation of steady-state processes including idling and throttling. Furthermore, they developed a transient process model through dynamic response fitting, ultimately constructing an integrated dynamic engine model by fusing transient and steady-state components. Liu [20] trained the MBGD network with system input, performance model output, and deviation data between the model and the real state of the engine. As shown in Fig. 3, a neural network-based performance model deviation correction framework was implemented, leading to enhancement of model-based control strategy through compensation for the discrepancy between the performance model and the engine's actual operating state. The improved controller significantly improves tracking accuracy within a deviation range of 10%, with no time delay or overshoot, almost no tracking error, and the same computational cost, thus greatly reducing the sensitivity to model mismatch; however, when the degree of model mismatch exceeds 10%, its suppression ability is still insufficient. Liu [20] used reinforcement learning methods to correct the deviation between the performance model and the real state of the engine, as shown in Fig. 4, to address the limited generalization ability of the improved neural network controller under model mismatch. The traditional model-based control method was improved, the output of the improved prediction model was used as the prediction model of the controller, and the cost function was reconstructed to establish an improved control method. The simulation results show that the improved controller can effectively control the tracking error within 2% without significantly increasing the consumption of computing resources, thereby significantly reducing the sensitivity to performance model mismatch, greatly improving the generalization ability under model mismatch, and effectively correcting model mismatch within a 30% state quantity deviation range. Cao [21] reformulated the simulation model calibration task as a deviation minimization problem between measured and simulated parameters, leveraging an adaptive cubic-mutation differential evolution algorithm to optimize the error function. The correction process for the simulation model is shown in Fig. 5. By using the test data of the aircraft engine, the twin-shaft mixed-flow turbofan engine simulation model was corrected. The correction results show that the deviation between the parameters of the corrected simulation model and the test parameters was reduced to within 1.0%, and the thrust, fuel consumption rate, and common working points of each component obtained by the engine simulation model all matched the test data, thus meeting the needs of engineering applications. Chen [22] developed a physics-based component-level representation of the DGEN380 turbofan engine, leveraging experimental data and multi-algorithm optimization (genetic algorithms, differential evolution, particle swarm optimization) to refine component characteristics through high-dimensional parameter space calibration. The research results show that the improved particle swarm algorithm was used to correct and update the component characteristic map of the DGEN380 at the design point, and the strategy of making small adjustments to the entire component characteristic map was adopted for the correction of the design point. The simulation results show that the relative error of the output target parameters of the corrected model meets the 1% accuracy requirement, thus achieving precise matching of the model design points.

Application of the Digital Twin Aircraft Engine Simulation Model

The applications of the digital twin aircraft engine simulation model in the field of aircraft engines can be divided into the following categories: performance optimization and prediction, fault diagnosis and prevention, and lifecycle management and maintenance optimization. Each category encompasses different technical applications, and specific cases are presented to better demonstrate how this technology promotes intelligence, precision, and efficiency in the aviation field.

Performance Optimization and Prediction

The application of digital twin technology in the field of aircraft engines, through real-time data collection and simulation analysis, assists airlines in optimizing engine

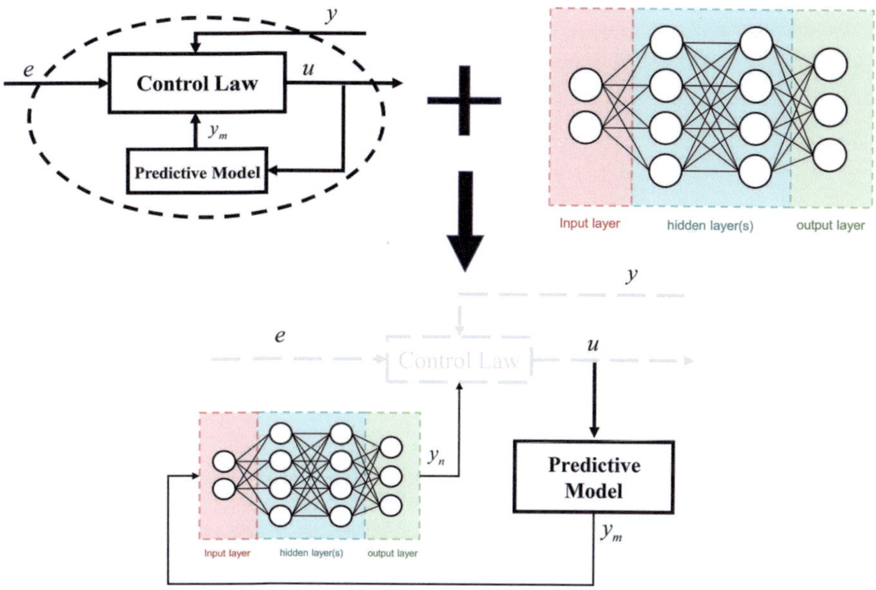

Fig. 3 Improved MPC controller based on the MBGD network correction method [20]

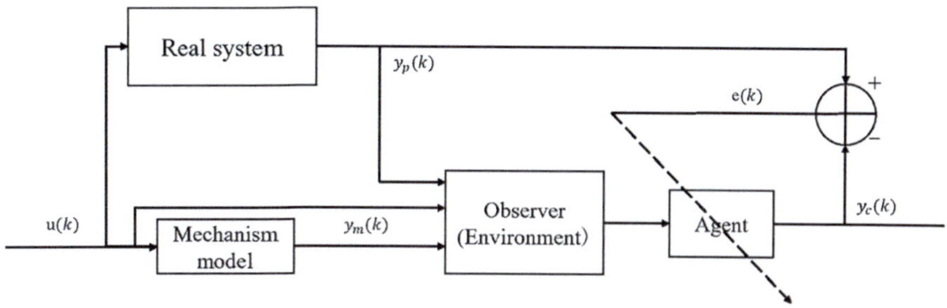

Fig. 4 Schematic diagram of the performance model correction method based on deep reinforcement learning

performance, particularly regarding fuel efficiency, thrust output, and performance under different flight conditions.

GE Aviation's GEnx engine, by leveraging digital twin technology, aims to improve fuel efficiency and thrust output precisely. As the core power system of the Boeing 787 Dreamliner, the GEnx engine significantly improves overall efficiency and maintainability by real-time monitoring and adjusting its digital twin model. This model continuously acquires flight data, dynamically updates and accurately predicts key performance indicators (KPIs) such as thrust and fuel consumption under different flight conditions. This real-time performance forecasting is vital to

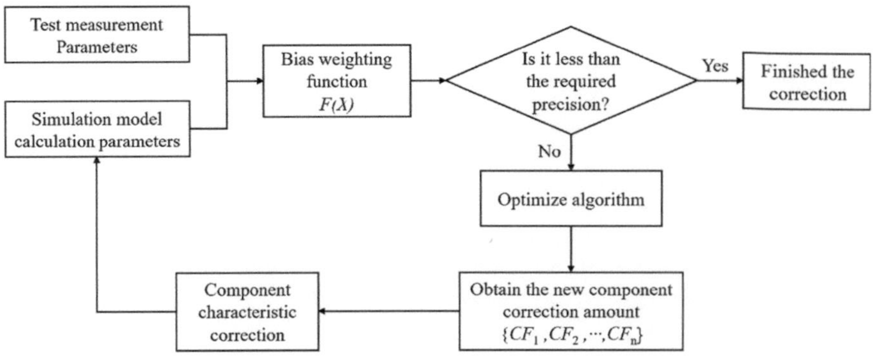

Fig. 5 Simulation model correction process

airlines because it can be used to optimize flight plans, reduce fuel consumption, and maximize range and fuel utilization.

By comparing digital twin models, GE Aviation further optimized engine design to ensure optimal performance under various conditions. These optimizations directly reduce operating costs and improve overall efficiency. On the basis of an analysis of more than 10,000 h of flight data, GE Aviation successfully predicted and improved the fuel efficiency of the engine. For example, in transcontinental flights from New York to Los Angeles, adjustments using digital twin models led to a 2 to 3% reduction in fuel consumption during flight, resulted in annual cost savings of approximately $500,000 in fuel costs per year on the basis of 2000 flights per year.

The Airbus A350 uses digital twin technology combined with flight data and an engine health monitoring (EHM) system to optimize performance during flight. This system can monitor engine status in real time, provide accurate performance data, and help airlines adjust flight plans in time to reduce fuel consumption and improve efficiency. Airbus's digital twin model can also make predictions on the basis of historical data and flight conditions to further optimize flight routes and aircraft scheduling and reduce unnecessary consumption. According to the digital twin data of the Airbus A350, airlines can save an average of 3–5% of fuel during flight each year. For large airlines, the annual savings in fuel costs can reach millions of dollars.

Fault Diagnosis and Prevention

The application of digital twin technology in fault diagnosis primarily involves real-time monitoring of engine health, analyzing data changes, predicting and warning of potential faults, and preventing the occurrence of sudden faults.

Rolls–Royce has deeply integrated digital twin technology with its flight data monitoring and health management system (FDM/HEMS) to construct an efficient engine health monitoring platform. By analyzing real-time data from the engine, the digital twin can simulate operational behavior and identify potential failure modes. This technology supports Rolls–Royce in maintenance optimization and enables predictive maintenance by airlines predict potential faults in the future.

The Trent XWB engine integrates digital twin technology. With this model, airlines can monitor the health status of the engine in real time, detect signs of wear or failure of turbine or compressor components in advance, and implement preventive maintenance to avoid accidental failures during flight. In 2024, the digital twin technology of Trent XWB successfully predicted approximately 100 potential faults, including turbine blade wear and incomplete combustion. For example, during an inspection, the digital twin system of Trent XWB detected an abnormal turbine blade temperature deviation (5 °C above nominal), and the system automatically issued a fault warning recommending maintenance during the next flight interval to avoid potential flight safety risks. Through these warnings, airlines avoided 7 engine failures caused by turbine damage, thus saving over $1.5 million in repair costs.

On the basis of long-term accumulated operating data, Rolls–Royce uses digital twin technology to analyze the operating status of turbines and compressors in detail. These data can help companies identify potential fault modes of engines under specific altitude and climate conditions. For example, in a hot and humid environment, the engine compressor may have airflow problems. The digital twin model, combined with historical data, can accurately predict the probability of such faults before the flight takes off and arrange maintenance in advance. Eight compressor faults caused by hot weather were successfully avoided. With the warnings and optimized maintenance from the digital twin model, the occurrence rate of sudden failures was reduced by about 20%. The average cost per repair event is approximately $100,000 (including aircraft downtime, repair, and logistics costs). With the preventive maintenance of digital twin technology, airlines have saved $5 million in repair expenses.

Lifecycle Management and Maintenance Optimization

Digital twin technology provides a new solution for the lifecycle management of aviation engines. Through real-time monitoring and data analysis, airlines can more accurately schedule the maintenance cycle of an engine, thereby extending the life of the engine, optimizing the allocation of maintenance resources, and reducing overall maintenance costs.

Pratt & Whitney's geared turbofan (GTF) engine manages its entire lifecycle through digital twin technology, thus helping airlines optimize maintenance plans and maximize the extension of engine life.

The GTF engine, which is equipped with digital twin technology, estimates the remaining useful life (RUL) of engine components on the basis of real-time data. For example, by monitoring the temperature, speed, and pressure of the engine's turbine blades, the GTF can predict the fatigue life of the turbine blades in advance, with wear estimated to occur after 5000 flight hours. With these data, Pratt & Whitney provide airlines with customized maintenance plans to ensure that the turbine blades are replaced prior to significant wear onset. According to the data analysis, the maintenance optimization of the GTF engine has advanced the replacement cycle of the turbine blades by 5–10%, thereby effectively avoiding sudden component failures, reducing flight downtime by 100 h, and saving an average annual maintenance cost of approximately 2 million dollars.

Leveraging the GTF digital twin dataset, predictive component maintenance and replacement strategies can be executed based on real-time engine health assessments, effectively eliminating the over-maintenance inherent in traditional scheduled maintenance protocols. For example, the digital twin model of the GTF clearly shows that during low-load conditions, components of the compressor and combustion chamber exhibit extended life spans, whereas they need to be replaced in advance under high-load conditions. This avoids major failures caused by excessive wear. According to actual data, on the basis of this precise prediction, the lifespan of the GTF engine has been extended by approximately 15%, resulting in approximately 10% savings in total maintenance expenditure, which is equivalent to saving about 30,000 dollars per flight.

Rolls–Royce uses digital twin technology extensively for lifecycle management in its Trent series engines. By collecting and analyzing large amounts of operating data, Rolls–Royce can track the usage status of an engine in real time and predict future maintenance needs. For example, the digital twin model of the Trent 700 engine combines the engine's operating data, environmental data, and past maintenance history and can accurately determine which components need maintenance or replacement, thereby maximizing the lifespan of the engine. In addition, Rolls–Royce can provide customers with condition-based customized maintenance schedules on the basis of these data, thereby helping airlines perform maintenance at appropriate times to ensure that engines runs in the best condition and avoid expensive unexpected downtime and maintenance costs.

We can see the wide application and significant advantages of digital twin technology in the field of aviation engines. Digital twins not only help airlines optimize engine performance and improve fuel efficiency but also play important roles in fault diagnosis, maintenance optimization, and lifecycle management. With the further development of digital twin technology, its application in the field of aviation engines will be more extensive, which will result in more efficient, intelligent, and low-cost operation modes for airlines.

Demand and Future Challenges of Aviation Engine Simulation Modeling

Aviation engine simulation modeling is a core technology in modern aviation engineering and provides strong support for design, optimization, and fault diagnosis. With the continuous advancement of aviation technology and the diversification of application requirements, simulation modeling in the field of aviation engines is becoming more important. Simulation technology can not only help engineers predict engine performance at the design stage but also provide dynamic monitoring during actual operation, thus helping solve complex fault diagnosis problems. However, with the continuous improvement in flight performance, the increasingly complex and dynamic operating environments, and the increasing requirements for engine operation efficiency and safety, aviation engine simulation modeling is facing a series of new demands and challenges.

Requirements of Aviation Engine Simulation Modeling

Complex Condition Modeling

Aviation engines experience various complex conditions in actual operation, such as transonic flow, high-temperature high-pressure combustion, and rotor dynamics coupling. The modeling of these conditions requires a deep understanding of the interactions among multiple disciplines, such as fluid mechanics, thermodynamics, and structural mechanics. For example, transonic flow involves situations where the airflow speed is close to the speed of sound, thus requiring the model to accurately describe the compression and expansion process of the airflow. For high-temperature high-pressure combustion, complex phenomena such as fuel and air mixing, combustion reactions, and heat transfer need to be considered. Rotor dynamics coupling focuses on the interaction between the rotor and other components, which affects the stability and performance of the engine. The modeling of these complex conditions place more stringent demands on simulation technologies.

Real-Time Requirements

During the flight process, an aviation engine needs to respond rapidly respond to in-flight contingencies. For example, a surge warning systems must process input data and issue alerts within milliseconds to ensure flight safety. This poses a severe challenge to the real-time computing capability of the simulation model. Traditional simulation methods may not meet such high real-time requirements, so it is necessary to develop more efficient algorithms and optimize model structures to achieve rapid calculations and responses.

Full Lifecycle Optimization

The performance prediction of aero engines is not limited to only the design phase but should also span the entire lifecycle, encompassing fault detection and degradation forecasting during the operation and maintenance phases. This requires that the simulation model be able to handle data from all stages from design to retirement and continuously update and refine. For example, during the operation and maintenance phase, the model needs to adjust parameters on the basis of actual operating data to reflect the current status and performance of the engine. In addition, the fault diagnosis model should have the ability to identify potential fault modes and provide early warnings to ensure timely maintenance and repair.

Future Challenges of Aero Engine Simulation Modeling

Traditional physics-based models heavily depend on expert heuristics and have difficulty fully capturing complex nonlinear dynamic processes. For example, the expansion of blade microcrack propagation is a highly nonlinear phenomenon, and traditional models often cannot accurately predict its development trend. In addition, the cost of experiments is high, and the single engine bench test can cost upwards of $1 million, which limits the practical application of traditional methods that rely on experimental verification. The computational inefficiency of multidisciplinary coupled simulations is also a major challenge. The iterative calculations in different disciplines, such as aerodynamics, thermodynamics, and structure, usually take a long time, which affects the overall research and development progress.

Data Barriers

As the aviation industry has entered the digital and intelligent era, the amount of operating data for aero engines has grown exponentially. Data from different types of sensors, detection equipment, and various monitoring systems present diverse characteristics in terms of sources, are usually distributed across disparate systems and systems, and lack standardized interfaces and interoperable data sharing protocols. This not only increases the complexity of data integration but also greatly limits the full utilization of data.

The data heterogeneity is also a key challenge. The operating data of aero engines include various physical quantities, such as temperature, pressure, speed, and combustion efficiency, and these data are often collected by sensors from heterogeneous vendors. The data formats, units, and transmission protocols may differ among sensors. This heterogeneous data type makes direct communication between different systems complex and inefficient. Therefore, the unification and processing of data from different devices are key steps in promoting data sharing and effective analysis.

Cross-platform data sharing is also an urgent problem to be solved. Many aero engine data storage platforms are independent of each other and are managed by different airlines, maintenance manufacturers, or research institutions. These platforms are usually based on different technical architectures or data storage methods, which makes data migration, access, and sharing between different platforms difficult. In addition, data security and privacy concerns must be addressed when sharing across platforms, especially when dealing with sensitive engine performance data; ensuring data security and confidentiality is particularly important.

To solve these problems, the aviation industry urgently needs to establish unified data standards. By unifying data formats, data exchange protocols, and data storage norms, the interoperability and accessibility of data can be significantly improved. For example, the use of open-standard data formats (such as JSON, XML, and CSV) and standard communication protocols such as RESTful APIs and MQTT can facilitate data interaction between different systems and platforms. In addition, the development of a cross-platform data sharing mechanism depends on the establishment of a unified standard interface and data exchange framework. In this way, data from different sensors and monitoring systems can be integrated seamlessly to form a cross-platform efficient data circulation system.

In addition, with the help of cloud computing and edge computing technologies, real-time processing and storage of data can be effectively supported. Centralized processing and analysis of massive amounts of engine data on the cloud platform can not only overcome physical limitations but also provide strong computing support for complex simulation modeling. Moreover, edge computing technology extends data processing to the edge close to the sensor, which reduces the data transmission delay, improves the response speed, and enables preliminary analysis to be conducted locally for timely feedback and decision-making.

The application of artificial intelligence and machine learning plays a crucial role in the integration of multisource data. Through deep learning and data fusion technology, hidden rules and associations can be identified in massive amounts of data, and data preprocessing, anomaly detection, pattern recognition, etc., can be performed to provide more accurate and reliable data support for simulation modeling. For example, the use of AI technology to automatically clean data, handle missing values and abnormal data, and reduce manual intervention can improve data quality. Moreover, machine learning models can automatically select the most suitable data fusion method on the basis of the characteristics of different data sources, thereby improving the accuracy and effectiveness of data analysis.

Real-Time Computing

To meet the real-time requirements of aero engine simulations and real-time warning systems, the development of lightweight models and efficient computing methods is particularly important. Traditional simulation models usually require many computing resources, especially when complex nonlinear dynamic systems are used, and the number of computations is large, which results in the model being unable to respond quickly during flight. Therefore, the simplification and optimization of existing models are the key to solving real-time problems.

Lightweight models achieve computational efficiency by minimizing model complexity and lowering model complexity. On the one hand, the number of parameters in deep learning models (such as neural networks) directly affects the consumption of computational resources. Therefore, simplifying the model structure and reducing the number of irrelevant or redundant parameters can effectively reduce the computational burden. For example, when convolutional neural networks (CNNs) or recurrent neural networks (RNNs) are used, reducing the number of network layers or reducing network depth or width can improve computational efficiency while ensuring accuracy. Moreover, the design of a model's input and output can also be optimized. By selecting more representative and important features as inputs, redundant data communication and computation can be reduced, thereby further accelerating the computation process.

In addition to model lightweighting, the efficiency of computational methods is crucial. Tools for accelerating inference, such as TensorRT, are typical examples of methods for increasing efficiency. TensorRT is a high-performance deep learning inference accelerator developed by NVIDIA and is specifically designed to accelerate neural network inference via GPUs. With TensorRT, trained deep learning models can be optimized through techniques such as model pruning, weight quantization, and layer fusion, thereby significantly improving inference speed and reducing computation time. In a real-time warning system for an aircraft engine, tools such as TensorRT can process massive amounts of data from sensors in a very short time, monitor the operation of the engine in real time, detect anomalies in a timely manner, and issue warnings to ensure flight safety.

In addition to hardware acceleration, optimizing algorithms is also an effective way to improve computational efficiency. Classic optimizers such as stochastic gradient descent (SGD) or Adam, can improve the convergence speed of a model to some extent, but there are still computational bottlenecks when dealing with large-scale data. In response to this problem, many new optimization algorithms, such as stochastic gradient descent (SGD) and the Adam optimizer, which can accelerate model training and reduce the time and computational resources needed, have emerged in recent years.

Multiphysics Coupling

In the multiphysics coupling analysis of aircraft engines, their operating characteristics are significantly affected by fluid–structure interaction (FSI), thermomechanical coupling, and nonlinear material behavior. Traditional decoupling strategies use sequential iterative methods to solve fluid motion equations, heat transfer equations, and structural dynamics equations separately. However, the discretization method, which is based on operator-splitting technique, faces numerical stability challenges such as ill-conditioned Jacobian matrices and cumulative errors in cross-scale parameter transfer. Especially in high Mach number regimes, the coupling effects of unsteady flow and thermoelastic deformation caused by shock waves and boundary layer interference can impair the convergence of traditional finite element-finite volume mixed hybrid solvers (FEM-FVM) by 40–60%.

The strong coupling solver based on the physics-informed neural network (PINN) employ a differential operator embedding framework, which integrates the Navier–Stokes equations, Fourier's law of heat conduction, and the Hooke–Kelvin viscoelastic constitutive equation as physical constraints embedded within the neural network architecture. This algorithm builds a composite loss function; performs multiobjective optimization of the control equation residuals, boundary condition constraints, and assimilation of measured data; and simultaneously resolves coupled flow, thermal, and structural fields. Compared with traditional explicit coupling methods, the PINN framework can reduce the data transfer error across the medium interface by two orders of magnitude while preserves spatiotemporal consistency and can effectively avoid numerical dissipation problems caused by the introduction of artificial viscosity terms.

Specifically, this architecture uses a deep network that combines adaptive activation functions and residual attention mechanisms and forms a physics-guided meta-learning paradigm by embedding physics-based constraints via differential operators. During the training process, the composite loss function is weighted by adjoint-based sensitivity analysis, and the intrinsic mode decomposition coefficients and network hyperparameters are optimized synchronously using the stochastic gradient descent algorithm. Experiments show that this method improves the solution efficiency of the aerothermal elasticity problem of turbine blades by 5.8 times compared with traditional CFD-CSD joint simulation, and the average relative error in high-gradient thermal field prediction remains below 1.2%, which verifies its engineering applicability in multiphysics strong coupling problems. Traditional numerical methods require additional processing and adjustment.

Trustworthy Artificial Intelligence (AI)

In the field of aircraft engines, with the widespread application of artificial intelligence (AI) technology, enhancing the interpretability and credibility of models has become key to ensuring safe, reliable, and efficient decision-making. Especially in critical application scenarios such as fault diagnosis and performance prediction, the "black box" nature of the model may lead to a lack of trust and understanding, which are critical for flight safety. Therefore, understanding and interpreting the decision-making processes of AI models in detail are important.

In the fault diagnosis process for aircraft engines, AI models predict potential fault modes by analyzing sensor data and historical fault records. However, given the complexity of aircraft engines, making decisions based solely on the predictions provided by a model is not enough. Airlines and maintenance engineers need to interpret the rationale behind model outputs and which factors play a key role in the prediction process.

The Shapley Additive exPlanations (SHAP) algorithm is an efficient interpretability tool that can analyze the prediction process of a model and reveal the effects of different input features on the results, thereby improving model interpretability, trustworthiness, and practical usability. With the help of the SHAP algorithm, the decision-making processes of AI models become more transparent. Engineers can not only know the prediction results of a model but also clearly quantify the contribution of each input variable on the performance prediction results. This helps engineers identify and verify key fault indicators, thereby enhancing the credibility of diagnostic results.

Metaverse Fusion

In the field of aircraft engine maintenance, with the continuous development of technology, the application of augmented reality (AR) and virtual reality (VR) technologies has become a critical approach to enhancing operational efficiency, safety, and maintenance effectiveness. Especially in equipment maintenance and troubleshooting links, AR and VR technologies provide technicians with an immersive and interactive experience but also effectively reduce the probability of human errors, increase the maintenance pace, and reduce the risks caused by improper operations. The fusion of these technologies has promoted the development of aircraft engine maintenance toward intelligence and digitalization, thus completely revolutionizing the traditional maintenance mode.

Virtual reality (VR) technology enables a fully immersive 3D simulation environment, which enables maintenance personnel to carry out operation training and troubleshoot without directly touching the aircraft engine. With the help of VR technology, maintenance personnel can practice various operations in a simulated environment and understand the equipment structure and operation process in detail, thereby reducing errors in actual maintenance. In a virtual scene, maintenance personnel can carry out troubleshooting training without pressure and without worrying about damaging the equipment or consuming too much time.

In aircraft engine training, VR technology can not only simulate various parts of the engine in detail but also reproduce the operating environment under different fault scenarios. This means that maintenance personnel can simulate high-risk scenarios in VR and carry out emergency handling and operation skills training; such scenarios may require complex preparations or high-risk tests in the real environment. Through this immersive training, maintenance personnel can more efficiently master maintenance skills, the training cycle for new employees can be shortened, and the overall technical level of the team can be improved.

AR overlays digital information onto the physical environment. Through AR glasses, smartphones, tablet computers or other smart devices, it projects virtual maintenance steps, operation procedures, and key component information into the field of view of maintenance personnel and displays operation steps and precautions in real time to help them to accurately complete maintenance tasks. For example, when facing complex engine disassembly tasks, the AR system can display the disassembly order, tool requirements, and related safety information through head-mounted display devices (such as smart glasses), thus reducing erroneous operations and repetitive labor.

In complex fault maintenance, airlines usually invite experienced experts to assist in remote diagnosis. With the help of AR technology, experts can remotely assist technicians via live video streaming, view real-time images of the maintenance site and interact with it. The maintenance personnel will aim the AR glasses at the parts to be repaired, and the expert will observe the real-time picture through the display screen, provide operation guidance, and even remotely mark the problem area through voice or graphic instructions to ensure the smooth progress of the maintenance process. This collaboration mode greatly improves the efficiency and accuracy of maintenance and plays an important role in cross-regional expert support and remote collaboration.

Conclusions

This article provides an in-depth exploration of the significance, historical development, modeling methodologies, practical applications, and future challenges of digital twin technology in aircraft engines in detail. This article traces the evolution of digital twin technology from conceptualization to practical maturity, which shows its rapid development and broad application potential in the aviation field. In terms of modeling methods, the article details the construction of performance mechanism models, data-driven adaptive model correction, modeling bias compensation mechanisms, and other innovative technologies. These methods, which combine multidisciplinary knowledge, real-time data, and optimization algorithms, significantly improve the accuracy and adaptability of the model, thus strongly supporting engine design, performance optimization, and fault diagnosis.

Through multiple actual application cases, specific outcomes of digital twin applications in optimizing engine performance, fault diagnosis, and life cycle management are presented. The article also discusses the requirements and challenges associated with aircraft engine simulation modeling, which include complex working condition modeling, real-time requirements, full life cycle optimization, data barriers, real-time computing, multiphysics field coupling, trustworthy AI, and metaverse fusion. Notwithstanding these challenges, the prospects for digital twin technologies in aero engine applications remain promising, and its continuous development will substantially enhance engine maintenance efficiency, operational safety, and cost-effectiveness.

In general, digital twin technology has become an indispensable key technology in the field of aircraft engines. Continuous technology optimization and innovation will provide more efficient and intelligent solutions for the full life cycle management of aircraft engines, thereby promoting the the intelligent and sustainable advancement of the aviation sector.

References

1. Yuan J, Liu Y, Gengfu L et al (2023) A review on mathematical modeling based on digital twin of aero engine. J Aerosp Sci Technol 11: 80
2. Song Y, Mou L, Liu Z et al (2022) Research on LPV modeling method of aero engine. In: China association for science and technology aero engine industry-academia consortium. proceedings of the sixth aerospace power joint conference and the 42nd technical exchange conference of the third professional information network of China aerospace and the 2021 aero engine technology development high-level forum vol V. School of Power and Energy, Northwestern Polytechnical University, p 7
3. Zheng P, Qi Y, Liu J et al (2023) Aero engine and gas turbine whole machine performance modeling method based on fluid network topology. Aero Engine 49(05):22–28
4. Yang K, Zong G, Wang W (2025) Aero engine component level identification modeling method based on hammerstein system. Journal of naval aeronautical university, pp 1-9, 18 Feb 2025
5. Gao Y, Wang X, Liu Z (2025) Research on aero engine performance modeling based on general simulation framework. In: Progress in aerospace engineering, pp 1–10 18 Feb 2025
6. Cheng C, Qiangang Z, Yong W et al (2023) Research on aero engine variable baseline model modeling method based on state awareness. Propuls Technol 44(09):196–210
7. Gao C (2022) Research on high precision modeling and starting performance simulation of aero engine Harbin Engineering University
8. Shuwei P, Qiuhong LI, Zhang H (2020) A new online modelling method for aircraft engine state space model. Chin J Aeronaut 33(6):1756–1773
9. Leibov R (2002) Aircraft turbofan engine linear model with uncertain eigenvalues. IEEE Trans Autom Control 47(8):1367–1369
10. Li C, Shungang N, Yang C et al (2023) On-wing aero engine performance digital twin modeling method. J Ordnance Equip Eng 44(06):204–212
11. Zheng J (2023) Research on digital twin modeling of aero piston engine. Civil Aviation Flight University of China
12. Haotian L, Dongwu G, Xunkai W et al (2024) Aero engine whole machine dynamics modeling and model verification. Propuls Technol 45(09):6–18
13. Liu B (2023) Research on modeling of micro turbojet engine power system based on test data. North China Aerospace Industry College
14. Lai R (2020) Prediction of remaining useful life of aero engine based on multi-source monitoring data fusion. Shanghai Jiao Tong University
15. Xiao L, Xu M, Chen Y et al (2019) Hybrid grey wolf optimization nonlinear model predictive control for aircraft engines based on an elastic BP neural network. Appl Sci 9(6):1254
16. Guo Y, Teng J, Zhou X et al (2024) Real-time adaptive model of mainstream parameters for aircraft engines based on OSELM-EKF. Aerosp Sci Technol 155:109662
17. Kim S, Kim K, Son C (2020) Transient system simulation for an aircraft engine using a data-driven model. Energy 196:117046
18. Pan M, Wang H, Huang J (2019) T-S fuzzy modeling for aircraft engines: the clustering and identification approach. Energies 12(17):3284
19. Li S, Peng J, Wang W et al (2022) Aero engine dynamic modeling based on flight parameter data and BP neural network. In: China association for science and technology aero engine industry-academia consortium. proceedings of the sixth aerospace power joint conference and the 42nd technical exchange conference of the third professional information network of China aerospace and the 2021 aero engine technology development high-level forum vol V. Air Force Engineering University, p 7
20. Liu C (2023) Research on advanced civil aero engine performance model deviation correction method based on intelligent optimization mechanism. Civil Aviation University of China
21. Mingdong C, Wei F, Bing L et al (2022) Aero engine simulation model correction method based on differential evolution algorithm. Gas Turbine Exp Res 35(02):1–6
22. Chen Y (2020) Research on aero engine model correction technology. Civil Aviation University of China

Enhanced Model of a Whole Aircraft Engine System

Abstract Aircraft engines involve multiple closely coupled disciplines. Owing to difficulties in bridging disciplines and the large workload of model interface adaptation, recent research has focused mainly on single-discipline and professional simulations, which have difficulty effectively reflecting the coupling relationships between various components and subsystems of aircraft engines. Therefore, this paper proposes an enhanced model of a whole aircraft engine system and provides a standard for evaluating the completeness of the model. On the basis of the engine performance model, a multisystem co-simulation model is formed by introducing detailed mechanism models of each subsystem. The digitally co-simulation of multisystem and multidisciplinary models can reveal the compatibility between subsystems in the design process in advance, mutually verify the rationality and completeness of the design results, and improve the success rate and credibility of the scheme design.

Keywords Aircraft engine · Enhanced model · Co-simulation · Coupling characteristics · Thermal coupling simulation

Introduction

The actual working process of an aircraft engine involves multiple disciplines, such as aerodynamics, thermodynamics, mechanics, hydraulics, and electronics, which are closely coupled. However, owing to difficulties in bridging professional tools and the large workload of model interface adaptation, most simulation technology research has focused mainly on single-discipline and professional simulations, such as "engine performance simulations" and "fuel system simulation". Effectively reflecting the coupling relationships between various components and subsystems of the aircraft engine is difficult, and effective verification and evaluation of the compatibility and mutual influence between subsystems and components are impossible. This leads to high costs and long cycles in the physical engine development process, as the design results and system performance can be adjusted only through trial and error [1].

Q. Xu et al., *Digital Intelligent Aeroengine*,
https://doi.org/10.1007/978-981-95-0942-3_3

In an actual physical system, the various components and subsystems of the engine are interrelated and work together to achieve operation of the engine. Therefore, digital models need the ability to simulate the integration of multiple system models on the basis of the original single-disciplinary modeling and simulation work to simulate the physical operating process and phenomena of the real engine. By digitally integrating multiple subsystems and professional models for simulation, the compatibility between subsystems in the design process can be discovered in advance to verify both the rationality and completeness of the design results and improve the success rate and credibility of the design scheme. Moreover, through simulation and integrated verification of the physical working mechanism, the operating process of a real engine under special conditions can be simulated and reproduced to predict and evaluate testing and operational risks.

In the 1990s, National Aeronautics and Space Administration (NASA) in America developed the numerical propulsion simulation system (NPSS), which supports engine component or whole engine modeling, simulations from zero to three dimensions [2–4], and multidisciplinary system coupling simulations [5]. In addition, NASA also launched the T-MATS toolbox [6], and the Air Force Research Laboratory (AFRL) built a turbo engine dynamic simulation system that supports dynamic coupling simulation verification among engine component-level real-time simulation models, control systems, and health management systems for real-time simulation verification of engine steady-state and transient processes.

The Central Institute of Aviation Motors in Russia developed the computer gas turbine engine simulator (CGTES), which mainly simulates the calculation of engine components or whole engine flow parameters and performance. It supports high-precision calculations of various dimensions, such as one-dimensional, S1, S2, and three-dimensional calculations, and can achieve coupled performance calculation analysis with air system [7].

In the Value Improvement Through a Virtual Aeronautical Collaborative Enterprise (VIVACE) plan [8], the propulsion object-oriented simulation software (PROOSIS) in the Virtual Engine project was developed in Europe. It focuses mainly on zero-dimensional and one-dimensional performance calculation simulations of an engine and supports coupled simulations of an aircraft and engine, as well as control, power, thermal management and other multidisciplinary co-simulation calculation functions. In addition to supporting traditional engine performance simulation, it can also support the modeling and simulation of aero hybrid power systems [9, 10].

China's aeroengine numerical simulation system (CANSS) supports the design optimization, simulation calculation, and process simulation functions for aero-engines. It can integrate third-party aerodynamic fluid calculation software and has the ability to simulate two-dimensional aerodynamics of a whole aero engine, such as the turbojet and turbofan, as well as S1, S2 and three-dimensional simulation capabilities [11, 12].

Although various countries have given attention to and invested in multidisciplinary modeling and simulation research on aero engines, the main focus has been on aerodynamic and thermal coupling modeling and simulation of engines. In terms of the whole system interconnection, the focus has been mainly on the co-simulation of control systems, health management systems and engines. The integrated simulation of other related subsystems (fuel systems, start-up power generation systems, lubricating oil systems, and air systems) still needs to be improved. The main reason is that it involves various professional software, and the model solvers that they rely on are different, which makes integrating the efficient and high-precision modeling and simulation functions of each profession in one software tool difficult, thus leading to serious model fragmentation problems.

To address the model fragmentation problem caused by multidisciplinary and cross-platform methods, Europe has proposed the functional mock-up interface (FMI) protocol standard [13] to support the co-simulation needs of multidisciplinary models, which has been widely promoted and applied. However, owing to the need for algorithm lightweighting and intellectual property protection in the model encapsulation process, the calculation accuracy and speed are affected after the model is integrated, thus making efficient integration difficult.

Although researchers have given attention to the importance and value of the integration and co-simulation of multidisciplinary and multisystem models of aero engines, owing to the involvement of many disciplines and professions in aero engines, there has been a lack of research on the integration methods and norms of models from different disciplines. Sorting and defining the integration schemes, solution methods and coupling degrees for the co-simulation of multiple systems of aero engines are necessary to better promote related research work.

System-Enhanced Model Definition and Evaluation Level

A system-enhanced model of an aero engine is a multisystem co-simulation model that is formed by introducing detailed mechanism models of each subsystem on the basis of the engine performance model, which can simulate the coupling and mutual influence between the engine and each subsystem. An actual physical engine requires the close cooperation of multiple subsystems to operate normally. In addition to the engine itself, the main subsystems include the control system, fuel system, lubricating oil system, air system, sensor system, start-up ignition system, gearing system, and power generation system.

Therefore, to establish a system-enhanced model of an aero engine, first establishing professional models of the engine and each subsystem that accurately reflect the working principles and processes of the engine and each subsystem is necessary. The engine performance model needs to simulate the response changes of the engine under the drive of control variables and external conditions on the basis of the characteristic parameters of the fan, compressor, combustion chamber, turbine, tail nozzle and other components of the aero engine. The control system model needs to simulate the main control functions, such as engine thrust management, start-up, steady state, acceleration and deceleration, limit protection, parking, surge suppression, fault diagnosis and fault-tolerant control, according to the designed control plan, control logic and control algorithm. The fuel system model needs to simulate the functions, performance, heating/exchange, fault simulation, etc., of the fuel system during engine operation through component-level modeling of the actual fuel regulator. The lubricating oil system model needs to simulate the main functions and requirements of the key components during the operation of the lubricating oil system, such as heating, heat exchange, and fault simulation, through component-level modeling of the engine lubricating oil system. The air system model needs to simulate the main functions of the air system, such as engine and key component bleed, heat transfer/cooling, pressure, and surge suppression. Similarly, for other subsystem models, mechanism models need to be established to realize the simulation calculation and modeling capabilities of the corresponding main functions.

After the professional models of each subsystem are established, the signal interactions between the engine performance model and other subsystem models should be done, make sure the model information interactions, data transmission and model simulation calculations under the same time base, the same conditions, and the same state. Before the engine performance model is calculated, the latest data, such as the control input of the fuel system model, the bleed and cooling/heat transfer data of the air system model, the status and input of the starter, the power consumption values of the gearing system model and the power generation system model, and the environmental conditions, need to be read. After the model parameters are calculated, the measured variables required by the sensor system model, the temperature and pressure parameters of the air system interaction interface, the bearing load and speed of the lubricating oil system, the speed of the gearing system, etc., need to be updated. The sensor system calculates the model output on the basis of the state parameters passed in by the engine performance model and sends it to the control system. The control system model needs to read the measurement results from the sensor system, accept the latest control instructions from the aircraft system, calculate the latest results according to the model, and update the control signals to parts such as the fuel system, starter, and power generation system. The fuel system model needs to read the latest values of the speed signal of the gearing system, the control signal of the control system, and the heat rise result of the lubricating oil system heat exchange; calculate the latest operation results according to the model mechanism; and send the updated control quantity to the engine performance model. The lubricating oil system model needs to read the latest results, such as the bearing load and speed of the engine, the heat exchange result for the fuel and lubricating oil, and the air pressure and flow in the bearing chamber; calculate the latest operation results according to the model mechanism; and update the parameters, such as the temperature and flow of the lubricating oil in the fuel and lubricating oil heat exchanger. The air system model needs to read the pressure and temperature values required from the engine performance model and update the bleed flow and cooling/heat transfer parameters at each position according to the model calculation results. Similarly, other subsystem models read the latest values of the input variables according to the input/output interaction relationship, and after the calculation is completed, the output results are updated to the latest values.

After establishing the engine aerodynamic and thermal performance model and its subsystem models and the interconnections between the models, establishing a comprehensive solution calculation scheme after the system parameters are coupled is necessary. The engine aerodynamic and thermal performance model and the models of the subsystems basically consider only their own professional knowledge and content in the modeling process. The parts that involve the interfaces and data interactions of each subsystem are mostly obtained by offline calculation and online interpolation to establish boundary conditions, and the simulation results of the internal independent solution calculation model are obtained. However, in the calculation of the enhanced model, a new coupling equation is formed in the process of parameter interaction of each subsystem model, which involves joint solution calculation for multiple model parameters or variables. The equations constructed in the traditional single model solution calculation scheme have difficulty reflecting the interactions and mutual influence between multiple systems, and constructing new solution equations for the engine performance model and each subsystem model is necessary for calculating the coupling between multiple system variables. Therefore, on the basis of the solution calculation for each subsystem model, developing a solver for the coupling equations of multiple systems is necessary for meeting the simulation calculation requirements of the coupling variables of each subsystem model (Fig. 1).

To measure and evaluate the coupling degree of the enhanced model of the whole aero engine system, the following five-level evaluation criteria are adopted according to the coupling signal types between each subsystem model. Starting from the most basic coupling of the engine aerodynamic and thermal performance parameters, through the introduction of the coupling calculation of signals such as electronic control, mechanical hydraulics, and heat exchange, a whole machine multisystem simulation calculation model that can interact with the aircraft or the test bench model is finally formed. The evaluation levels shown in the table correspond to enhanced models that are formed on the basis of the model of the previous level by adding new coupling parameters. The modeling requirements for the models of various levels are as follows.

The L0-level model is a steady-state parameter model of the aero engine that is composed of steady-state models of each subsystem and is used to calculate the matching of the performance parameters of each steady-state point of the engine.

The L1-level model is an engine aerodynamic and thermal performance model that calculates the dynamic responses of the aero engine from the input, such as the rotor speed, sectional parameters (temperature, pressure, flow, etc.), thrust, and simulates the engine starting, acceleration and deceleration, steady state, parking, etc.

The L2-level model is based on the L1-level model, with the addition of a coupling simulation function of electronic control signals, and is used to calculate the response of the closed-loop system, which is composed of the engine model, control model, sensor model, etc., under external commands and needs to include engine control plans, control logic, control algorithms, etc., and can support the simulation capabilities for sensor performance characteristics and faults.

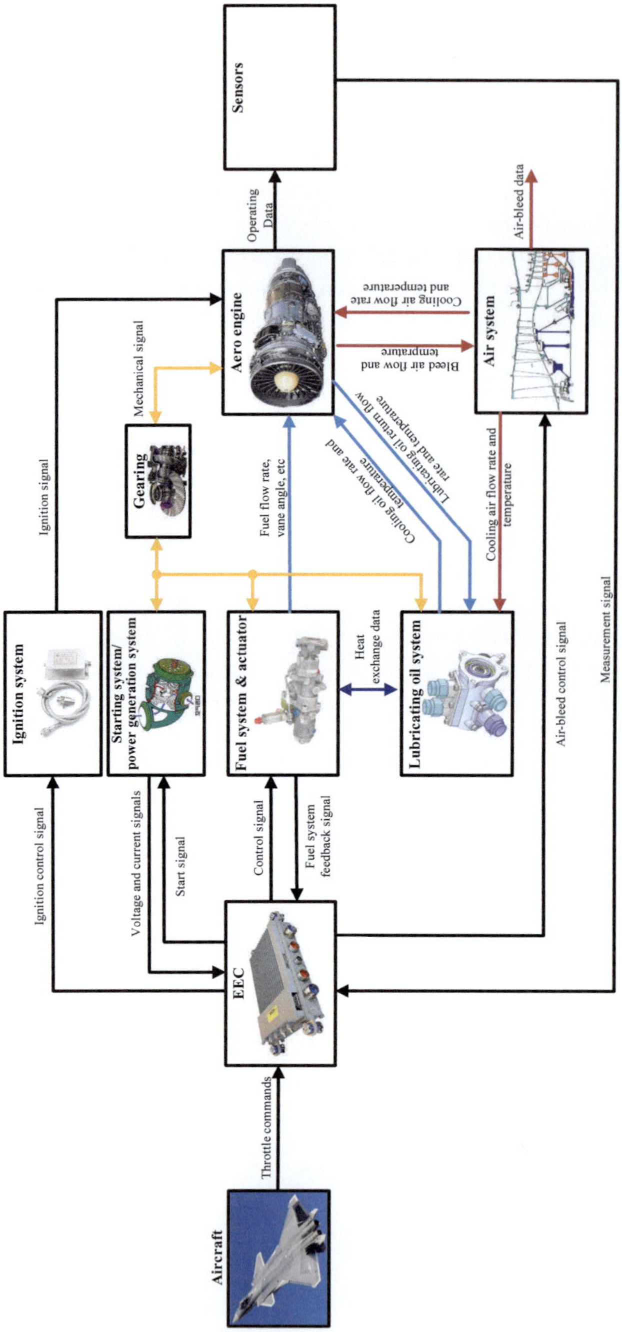

Fig. 1 Schematic diagram of the whole aero engine system

The L3-level model is based on the L2-level model. The coupling simulation function of the mechanical hydraulic parameters is added to calculate the system response results when the component-level mechanism model of the mechanical hydraulic system is considered, and the fuel, gearing, starter, generator, load and other system component-level mechanism models need to be established to support the simulation capabilities for the mechanical hydraulic performance characteristics and faults.

The L4-level model is based on the L3-level model, with the addition of the coupling simulation function of heat exchange parameters, which is used to calculate the heat exchange and temperature rise conditions between different systems of the aero engine with changes in engine working state. Component-level mechanism models of the oil system, air system and related systems (such as a fuel system thermal simulation model and fuel–oil heat exchange model) need to be established, which can support the simulation capabilities for heat exchange performance characteristics and faults.

The L5-level model is based on the L4-level model; additionally, a coupling simulation function of external condition parameters is used to calculate the response of the aeroengine system model coupled with the aircraft or test bench model, and the coupling of mechanical, electrical, control, gas, liquid, heat and other signals between the aeroengine model and the aircraft or test bench model needs to be realized, which can support the simulation capabilities for the dynamic evolution of environmental conditions.

In the process of building the enhanced model of the whole aero engine system, in the early stage of low-level modeling simulation, the high-level subsystem models involved can be replaced by simplified models to ensure the normal simulation calculation function of the model. For example, in stage L1 or L2, the fuel model and starter model required in the simulation process can be replaced by simplified models or linear models to ensure the normal transmission and reasonable interaction of signals in the model simulation calculation process (Table 1).

Modeling and Solving

To illustrate the benefits of multidisciplinary co-simulation of aero engines in the design and development process, a turbofan engine is taken as an example. By establishing a co-simulation model of engine performance and the fuel, control, lubricating oil, air and other subsystems, the dynamic response coupling characteristics between the engine–fuel–control subsystems and the thermal coupling characteristics between the fuel/lubricating oil systems are analyzed. Figure 2 shows a modeling diagram of the aero engine verification case. The aircraft system model gives the control system a throttle command to control the engine operating state. The control system model performs closed-loop control according to the throttle lever command, control plan, control law and sensor collected data; obtains the fuel flow control quantity; calculates the voltage and current control quantity required by the fuel system according

Table 1 System enhancement model coupling degree level standard

Level	Coupling parameter	Involved systems	Function
L0	Steady-state parameter coupling	Steady-state models of each subsystem	Calculates the matching of performance at each steady-state point
L1	Aerodynamic and thermal performance parameter coupling	Engine performance model Air system 0D characteristic model	Supports engine start-up, acceleration/ deceleration, steady state, parking and other process simulation capabilities and can simulate changes in key engine performance parameters with input
L2	Electronic signal coupling	Control plan, control logic and control algorithm Sensor model Ignition system model	Includes at least an engine control plan, control logic, control algorithm, etc., and supports sensor performance characteristics and fault simulation capabilities
L3	Mechanical hydraulic parameter coupling	Fuel system Starter Generator Gearing system Load system	Fuel system, gearing, starter, generator, load and other systems reach component-level mechanism simulation and support performance characteristics and fault simulation capabilities
L4	Heat exchange parameter coupling	Lubricating oil system model Air system 1D model	Lubricating oil, air and other system models reach component-level mechanism simulation and support performance characteristics and fault simulation capabilities
L5	Environmental condition coupling	Platform interaction interface Inlet and exhaust conditions Special working conditions	Realize the coupling of mechanical, electrical, control, pneumatic, hydraulic, thermal and other signals between the engine and other subsystems, and support dynamic evolution simulation capability that varies with environmental conditions

to the fuel regulation control law; and inputs the results into the fuel system model. The fuel system model calculates the fuel pump speed according to the engine speed and gearbox characteristics, obtains the desired fuel flow according to the external given control quantity, and inputs it into the turbofan engine model after the engine operating state. The fuel system model then calculates the heat exchange data through the fuel lubricating oil heat exchanger; calculates the next-moment engine operating parameters according to the fuel flow and guide vane control angle, as well as the current engine operating state; sends the results into the sensor model; and sends the processed data as feedback data into the control system model. The lubricating oil system model calculates the lubricating oil temperature according to the engine operating state and then calculates the heat exchange data through the fuel lubricating oil heat exchanger. The air system model calculates the bleed air quantity that is input

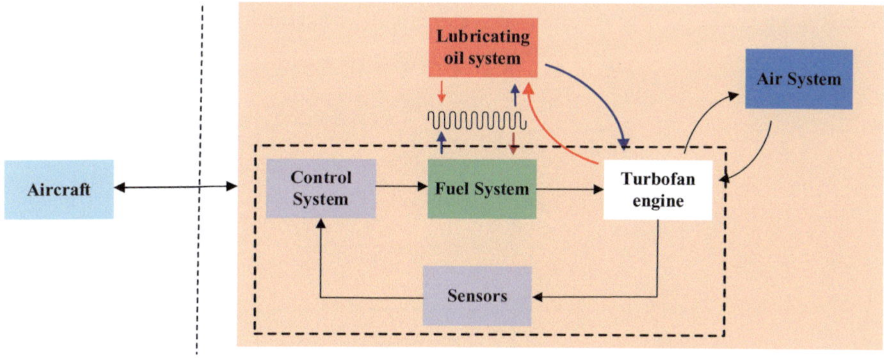

Fig. 2 Co-simulation framework

into the engine model according to the temperatures and pressures at the inlet and outlet (Fig. 2).

Engine Performance Modeling

Analysis of the dynamic characteristics of aeroengines is a cross-disciplinary field that integrates mechanics, fluid mechanics and thermodynamics. According to the first law of physics, three sets of basic dynamic equations can be used to describe the operation of aero engines: rotor dynamics caused by inertial effects, pressure dynamics caused by mass storage effects, and temperature dynamics caused by energy storage effects [14]. The proposed mathematical model of an aero engine essentially applies these effects to a complete engine from inlet to nozzle. This section takes the turbofan engine as an example to introduce its modeling process.

Figure 3 shows a diagram of the basic principles of turbofan engine modeling, which covers the main components of the turbofan engine model, including the inlet, low-pressure compressor, high-pressure compressor, combustion chamber, high-pressure turbine, low-pressure turbine, bypass duct, mixing chamber, and nozzle. This diagram also includes the flow and pressure balance equations that need to be satisfied on the basis of the matching relationship between each component, as well as the rotor dynamics equations that need to be satisfied by the changes in the rotational speeds of the high- and low-pressure rotors. These balance equations and rotor dynamics equations together constitute the common working equations of the aero engine for the turbofan engine, which include the following common main working equations:

Note: The engine section flow W and the power of the rotating parts *Torque* are both nonlinear functions of nH, nL, B, π_{HPC}, π_{LPC}, π_{HPT} and π_{LPT}. The specific nonlinear relationship is determined by the mathematical model and the characteristics of the engine components.

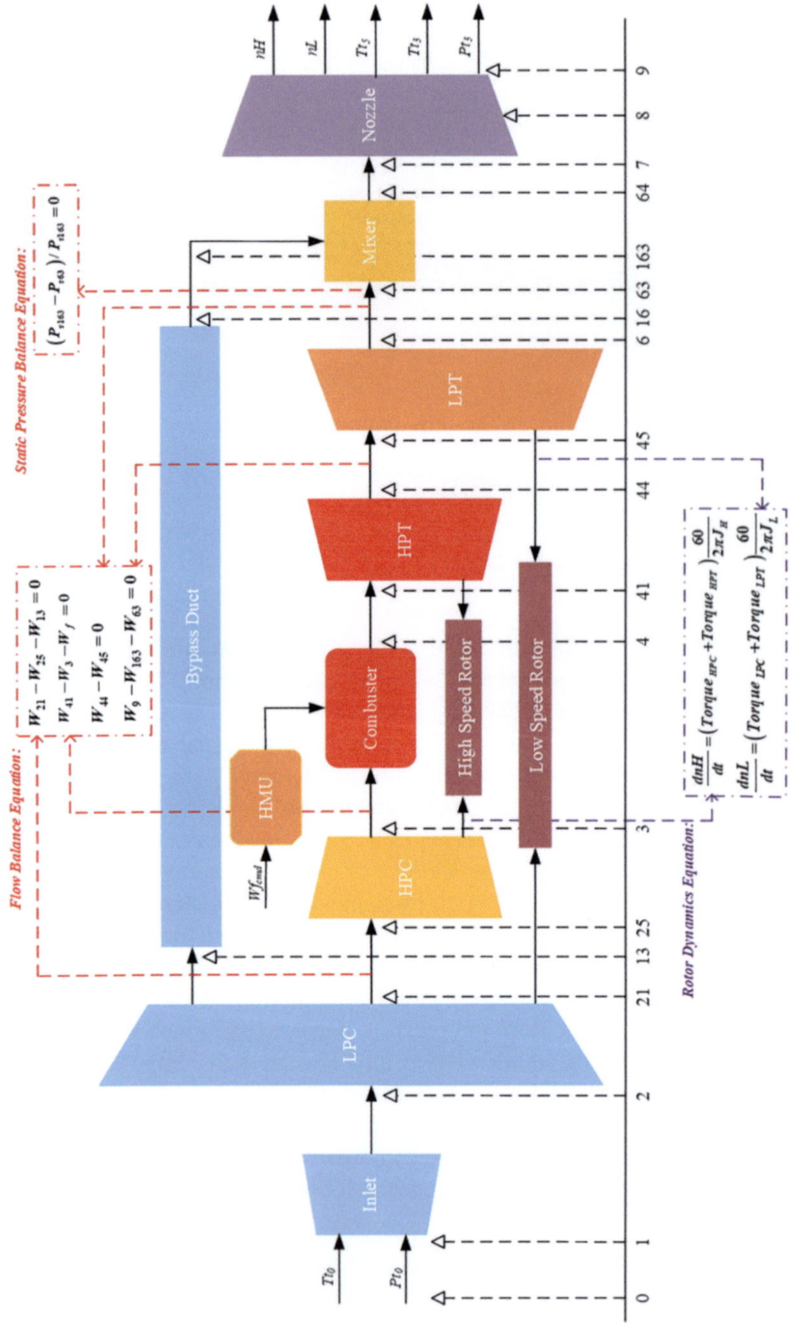

Fig. 3 Basic principles of turbofan engine modeling

1. The balance equation between the outlet flow of the low-pressure compressor and the sum of the inlet flows of the high-pressure compressor and bypass is as follows:

$$f_1(nH, nL, B, \pi_{HPC}, \pi_{LPC}, \pi_{HPT}, \pi_{LPT}) = W_{21} - W_{25} - W_{13} = 0 \quad (1)$$

where W_{21} represents the outlet flow of the low-pressure compressor, W_{25} represents the inlet flow of the high-pressure compressor, and W_{13} represents the inlet flow of the bypass duct.
2. The balance equation between the inlet flow of the high-pressure turbine and the sum of the outlet flow of the compressor and the fuel quantity is as follows:

$$f_2(nH, nL, B, \pi_{HPC}, \pi_{LPC}, \pi_{HPT}, \pi_{LPT}) = W_{41} - W_3 - Wf = 0 \quad (2)$$

where W_{41} represents the inlet flow of the high-pressure turbine, W_3 represents the outlet flow of the compressor, and Wf represents the fuel flow.
3. The balance equation between the inlet flow of the low-pressure turbine and the outlet flow of the high-pressure turbine is as follows:

$$f_3(nH, nL, B, \pi_{HPC}, \pi_{LPC}, \pi_{HPT}, \pi_{LPT}) = W_{44} - W_{45} = 0 \quad (3)$$

where W_{44} represents the outlet flow of the high-pressure turbine and W_{45} represents the inlet flow of the low-pressure turbine.
4. The balance equation between the outlet flow of the nozzle and the sum of the outlet flows of the bypass and low-pressure turbine is as follows:

$$f_4(nH, nL, B, \pi_{HPC}, \pi_{LPC}, \pi_{HPT}, \pi_{LPT}) = W_9 - W_{163} - W_{63} = 0 \quad (4)$$

where W_9 represents the outlet flow of the nozzle, W_{163} represents the outlet flow of the bypass duct, and W_{63} represents the outlet flow of the low-pressure turbine.
5. The static pressure balance equation of the mixing chamber is as follows:

$$f_5(nH, nL, B, \pi_{HPC}, \pi_{LPC}, \pi_{HPT}, \pi_{LPT}) = (P_{s163} - P_{s63})/P_{s163} = 0 \quad (5)$$

where P_{s163} represents the static pressure at the outlet of the bypass duct and P_{s63} represents the static pressure at the outlet of the low-pressure turbine.
6. The power dynamic equation of the high-pressure compressor and high-pressure turbine is as follows:

$$f_6(nH, nL, B, \pi_{HPC}, \pi_{LPC}, \pi_{HPT}, \pi_{LPT}) = \frac{dnH}{dt}$$

$$= (Torque_{HPC} + Torque_{HPT})\frac{60}{2\pi J_H} \tag{6}$$

where nH represents the rotational speed of the high-pressure rotor, $Torque_{HPC}$ represents the output power of the high-pressure compressor, and $Torque_{HPT}$ represents the output power of the high-pressure turbine.

7. The power dynamic equation of the low-pressure compressor and low-pressure turbine is as follows:

$$\begin{aligned} f_7(nH, nL, B, \pi_{HPC}, \pi_{LPC}, \pi_{HPT}, \pi_{LPT}) &= \frac{dnL}{dt} \\ &= (Torque_{LPC} + Torque_{LPT})\frac{60}{2\pi J_L} \end{aligned} \tag{7}$$

where nL represents the rotational speed of the low-pressure rotor, $Torque_{LPC}$ represents the output power of the low-pressure compressor, and $Torque_{LPT}$ represents the output power of the low-pressure turbine.

Therefore, the engine model can be represented as a set of nonlinear equations determined by these balances and dynamic equations:

$$f_i(nH, nL, B, \pi_{HPC}, \pi_{LPC}, \pi_{HPT}, \pi_{LPT}) = 0,\ \ i = 1, \cdots, 7 \tag{8}$$

Given independent guess variables and control quantities, solving the model becomes a problem of solving nonlinear equations. Nonlinear equations are generally solved via numerical methods in mathematics. The most classic and common method is the Newton–Raphson (N–R) method. The N–R method solves nonlinear equations iteratively. First, a set of initial guess values are substituted into the equations to calculate a set of error values, and the error values are used to correct the initial guess values. Then, the values are recalculated until the absolute value of the error meets a certain accuracy requirement; when this requirement is satisfied, the solution of the equations is obtained.

We let $F(x) = [f_1, f_2, f_3, f_4, f_5, f_6, f_7]^T$ and $x = [nH, nL, B, \pi_{HPC}, \pi_{LPC}, \pi_{HPT}, \pi_{LPT}]^T$. The iterative formula for solving the above engine model nonlinear equations with the N–R method is as follows:

$$x^{k+1} = x^k - J^{-1}F\left(x^k\right)\ k = 0, 1, 2, \cdots \tag{9}$$

where J represents the Jacobian matrix and where x^k represents the vector of the solution at step k; if $k = 0$, x^k is the initial guess vector.

Fuel System Modeling

A principle diagram of the main fuel system is shown in Fig. 4, and the structure of the main fuel system is shown in Fig. 5, which includes 16 unknown variables, such as the pressure after fuel filtration p_p, the outlet pressure of the metering valve p_m, the outlet pressure of the parking valve p_h, the return fuel flow rate before the gear pump q_r, and the outlet flow rate of the metering valve q_m. There are 6 known quantities: the outlet pressure of the centrifugal pump p_i, the pressure after the nozzle p_b, the outlet flow rate of the gear pump q_p, the pressure at the P end of the electrohydraulic servo valve p_d, the low-pressure return fuel pressure p_t, and the input current of the electrohydraulic servo valve I_m. The first state quantity is the displacement of the metering valve core x.

The system description can be divided into a steady-state part and a dynamic part. The steady-state part consists of the connection relationships between the components, which can be represented by algebraic equations. According to the conservation of flow and the components, such as spray nozzles, metering valves, and constant pressure valves, a total of 16 algebraic equations can be obtained, which is consistent with the number of unknowns:

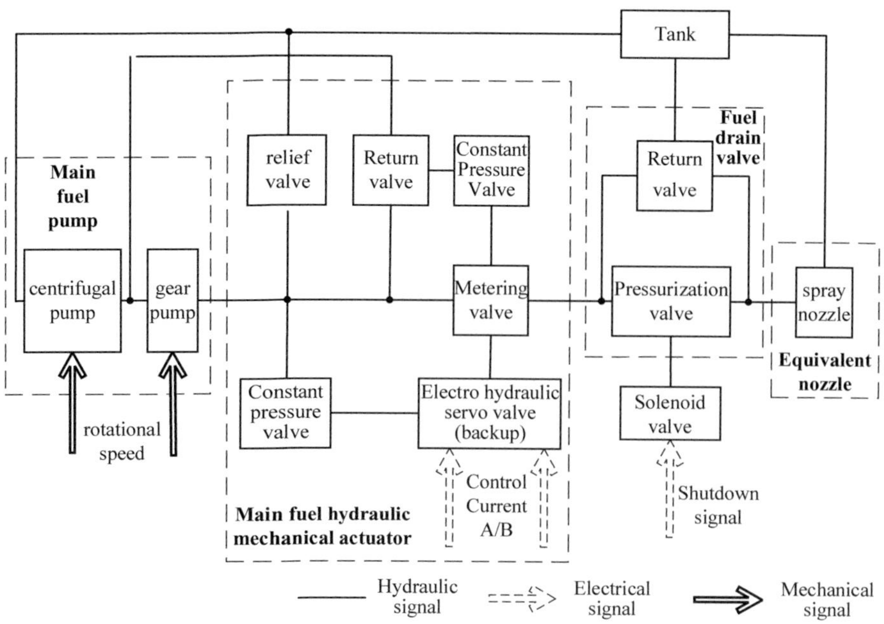

Fig. 4 Principle diagram of the main fuel system

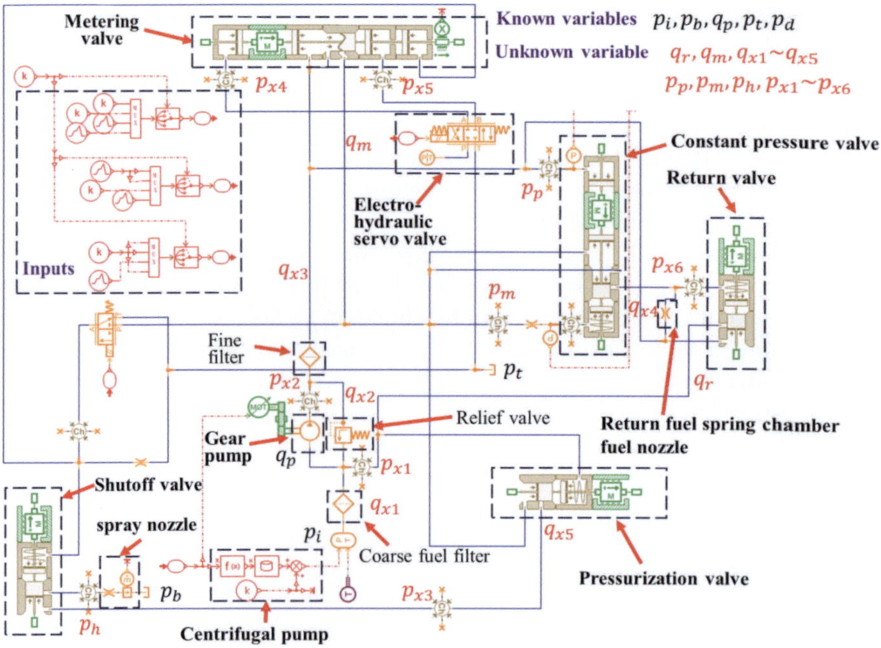

Fig. 5 Structure of the main fuel system

$$
\begin{cases}
p_h = f_{spraynozzle}(p_b, q_{x5}) \\
p_{x3} = f_{shutoffvalve}(p_h, p_t, q_{x5}) \\
p_{x1} = f_{Pressurization}(p_m, p_{x3}, q_{x5}) \\
q_{x1} = f_{Coarsefilter}(p_i, p_{x1}) \\
p_{x5} = f_{meteringvalve\_A}(p_{x4}) \\
p_{x5} = f_{servovalve}(p_d, p_t, p_{x4}, I_m) \\
q_m = f_{meteringvalve\_B}(p_m, p_p, x) \\
q_{x5} = q_{x4} + q_m \\
p_{x6} = f_{constantpressurevalve}(p_m, p_p, q_{x4}) \\
q_{x4} = f_{fuelnozzle}(p_{x6}, p_p) \\
q_r = f_{returnfule}(p_{x6}, p_p) \\
q_{x3} = q_{x5} + q_r \\
p_{x2} = f_{finefilter}(p_p, q_{x3}) \\
q_{x2} = f_{reliefvalve}(p_{x2}, p_{x1}) \\
q_p = q_{x1} + q_{x2} + q_r \\
q_p = q_{x2} + q_{x3}
\end{cases}
\tag{10}
$$

The dynamic description considers only the most important dynamics of the main fuel system, that is, the first-order motion equation of the metering valve core. According to the flow rate of the electrohydraulic servo valve and the size of the metering valve, the state equation of the displacement of the metering valve core is expressed by the following formula.

$$\dot{x} = f_{servo\&metering}(p_d, p_t, p_{x4}, I_m) \tag{11}$$

On the basis of the above system description, the system solution equations can also be divided into algebraic equations and differential equations. The algebraic equations can be solved by applying the Newton–Raphson algorithm from the basic function library, and the differential equations can be integrated directly via the first-order Euler method. Notably, when the convergence of the Newton–Raphson algorithm is not good, improved measures such as divide and conquer, local iteration, and binary solution can be used.

Control System Modeling

The turbofan engine control system includes mainly the main fuel flow control, adjustable blade angle control at the inlet of the low-pressure compressor, adjustable stator blade angle control at the inlet of the high-pressure compressor, afterburner first zone fuel flow control, afterburner internal combustion fuel flow control, afterburner outer annular fuel flow control, and nozzle A8 area control. For the control loops, hydraulic servo loops are commonly used, and the electronic controller controls the hydraulic servo loops through an electrohydraulic conversion device. The main fuel flow control is generally used to adjust the engine operating state and performance. Therefore, the main fuel flow control is used as an example to introduce the engine control system model building process.

The engine main fuel flow control generally consists of three parts: a control plan, a control algorithm, and fuel flow control. The engine control plan calculates the current thrust control target parameters of the engine on the basis of the control input and environmental conditions. The control algorithm, which generally includes acceleration/deceleration control and protection control functions, is responsible for achieving the specified target parameters. It should simultaneously guarantee the engine thrust adjustment control function and system operational safety. The fuel flow control adjusts the valve opening according to the fuel flow command generated by the engine control to adjust the engine fuel supply flow (Fig. 6).

For the implementation of the control plan, the thrust control target parameter values are obtained mainly on the basis of the current state and input parameters, where the speed or pressure is usually the main variable of the thrust adjustment control:

$$nH_{cmd} = f(Throttle, Tt2, Pt2, nH_{\max}) \tag{12}$$

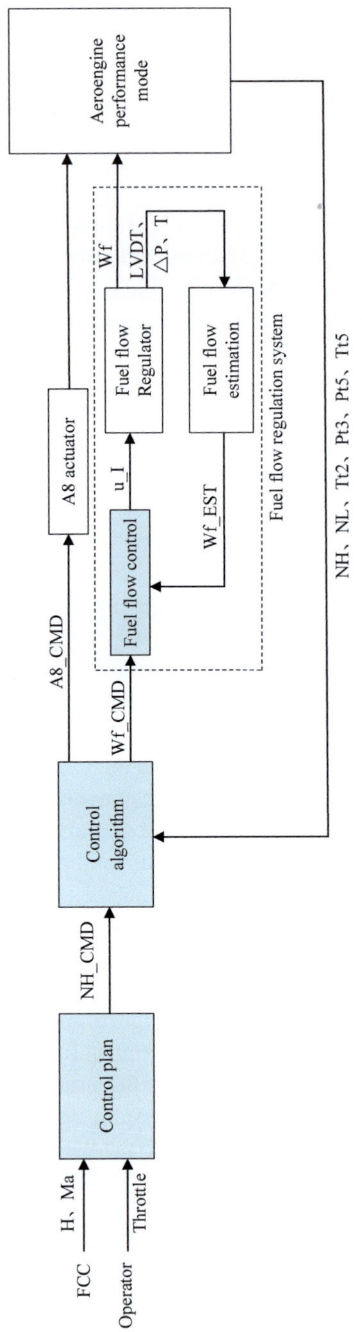

Fig. 6 Block diagram of the engine control scheme

In addition, obtaining the set values of the engine limit protection parameters is necessary:

$$\begin{aligned} nL_{\max} &= f(Tt_2) \\ nH_{\max} &= f(Tt_2) \\ Tt_5^{\max} &= f(Tt_2) \\ Pt_3^{\max} &= f(Tt_2) \end{aligned} \tag{13}$$

The engine controls the thrust of the engine by adjusting the speed/pressure ratio, so the speed/pressure ratio adjustment control is the core part of the engine control. However, in addition to ensuring engine control performance, avoiding entry into unsafe operating boundary areas during engine operation, which affects engine safety and service life, is also necessary. To integrate speed control and protection control, a switching control method that is based on min–max selection logic is adopted, which generally includes steady-state controllers, acceleration control modules, deceleration control modules, overtemperature protection modules, overspeed protection modules, overpressure protection modules and other control parts. According to the size of the control quantity of each module and the selection logic, a route is selected as the control output at the current time to adjust the engine speed and achieve a safe and stable closed-loop control effect (Fig. 7).

The control plan related to the main fuel control loop includes the following aspects.

Speed/pressure ratio control fuel flow:

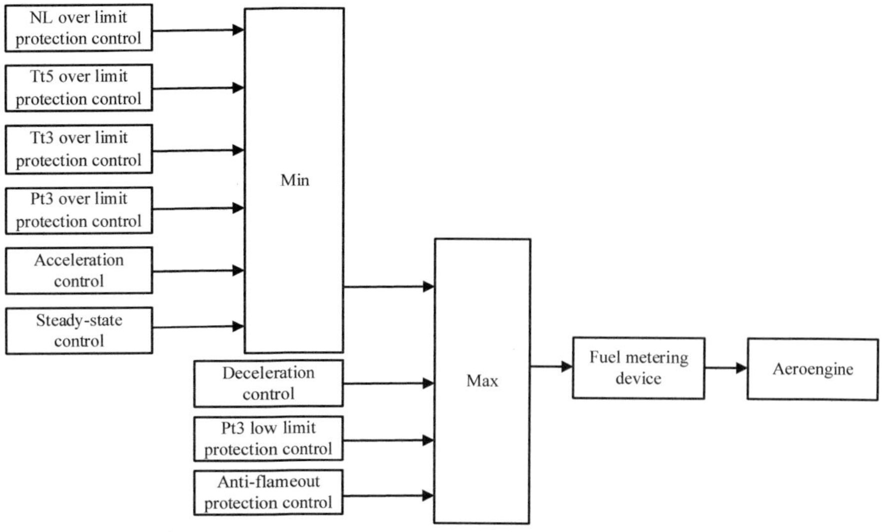

Fig. 7 Switching control scheme based on min–max selection logic [14]

$$\Delta Wf_{SS} = f(nH_{\text{cmd}}, nH_{\text{fbk}}) \tag{14}$$

Acceleration and deceleration fuel flow:

$$\Delta Wf_{acc}, \Delta Wf_{dec} = f(nH_C, Tt2, Pt3) \tag{15}$$

Limit protection control fuel flow:

$$\begin{aligned}\Delta Wf_{\text{limit}}^{\text{Pt3}} &= f(Pt3_{\max}, Pt3_{\text{fbk}})\\ \Delta Wf_{\text{limit}}^{Tt5} &= f(Tt5_{\max}, Tt5_{\text{fbk}})\\ \Delta Wf_{\text{limit}}^{nL} &= f(nL_{\max}, nL_{\text{fbk}})\\ \Delta Wf_{\text{limit}}^{Tt3} &= f(Tt3_{\max}, Tt3_{\text{fbk}})\end{aligned} \tag{16}$$

The main fuel flow of the engine is determined by multiple control loops, including a steady-state control loop, some parameter limit control loops, acceleration and deceleration control loops, and so on. The min–max selection logic determines which control loop should be active at any given time, thus achieving multiple control objectives at the same time.

$$\Delta Wf_{cmd} = \min\left(\max\begin{pmatrix}\Delta Wf_{SS}, \Delta Wf_{acc}, \Delta Wf_{\text{limit}}^{Pt3},\\ \Delta Wf_{\text{limit}}^{Tt5}, \Delta Wf_{\text{limit}}^{nL}, \Delta Wf_{\text{limit}}^{nH}\end{pmatrix}, \Delta Wf_{dec}\right) \tag{17}$$

$$Wf_{cmd} = f(Wf_{cmd}, Wf_{\max}, Wf_{mix}, Wf_{cmd},) \tag{18}$$

The fuel flow control part calculates the current control signal to the fuel adjustment device on the basis of the calculated fuel flow command, as well as the signal feedback from the fuel adjustment device (LVDT, ΔP, T_fuel, nH), etc.

$$u\_I = f\left(Wf_{cmd}, LVDT, \Delta P, T_{fuel}, nH\right) \tag{19}$$

Lubrication System Modeling

The aircraft engine lubrication system, which is an important auxiliary system of the engine, consists of oil supply, oil return, and ventilation subsystems. It is mainly responsible for lubricating and cooling the internal bearings, gears, splines, and other transmission parts of the engine to ensure the normal operation of the engine. On the basis of the law of conservation of energy and the principles of fluid mechanics, the operation of a lubrication system can be described by three sets of basic equations that

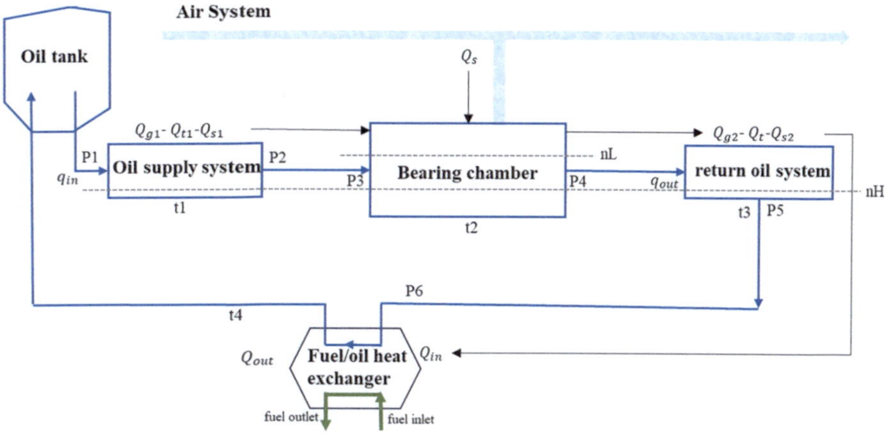

Fig. 8 Schematic diagram of the lubrication system structure

describe the following dynamics: fluid dynamics caused by flow changes, pressure dynamics caused by pressure changes, and thermodynamics caused by temperature changes. Here, a forward single-cycle closed lubrication system is introduced.

The structural principle of the positive single-loop lubrication system is shown in Fig. 8, which includes an oil supply system, bearing chamber, oil return system, fuel/oil cooler, oil tank, ventilation system, etc. This block diagram also contains the flow and pressure balance equations that need to be satisfied on the basis of the matching relationship between each component (Fig. 8).

The heat balance equation of the boosting system is as follows:

$$f_1(nH, \eta, p, q, \Delta t_1) = Q_{g1} = Q_{m1} + Q_{t1} + Q_{s1} \tag{20}$$

where Q_{g1} represents the heat from the friction and efficiency loss of the booster pump, Q_{m1} represents the heat lost during lubricating oil cooling, Q_{t1} represents the heat lost to the surrounding environment, and Q_{s1} represents the heat stored in the lubrication system.

The heat balance equation of the bearing chamber is as follows:

$$f_2(nH, nL, \rho, d, \mu, \Delta t_2 = Q_g + Q_s = Q_c + Q_r + Q_{l1}) \tag{21}$$

where Q_g represents the heat generated in the bearing chamber due to various heat sources, which include mainly the heat generated by friction between the bearing rolling body and the raceway and the heat generated by the shear of the lubricating oil viscosity, Q_s refers to the heat transferred into the bearing chamber by other means, such as the air system and heat conduction, Q_c represents the heat taken away by the lubricating oil, Q_r represents the heat lost to the surrounding environment through heat radiation, and Q_{l1} represents the heat stored in the parts at the bearing chamber.

The heat balance equation of the heat exchanger is as follows:

$$f_3(nH, \eta, p, q, \Delta t_3) = Q_{in} = Q_{out} + Q_{l2} \tag{22}$$

where Q_{in} represents the total heat carried by the lubricating oil when it enters the heat exchanger, Q_{out} represents the heat taken away by the fuel system, and Q_l represents the heat stored in the lubrication system.

The heat balance equation of the oil return system is as follows:

$$f_4(h, A, \Delta t_4) = Q_{g2} = Q_m + Q_t + Q_{s2} \tag{23}$$

where Q_{g2} represents the heat from the friction and efficiency loss of the return oil pump, Q_{m2} represents the heat lost during lubricating oil cooling, Q_{t2} represents the heat lost to the surrounding environment, and Q_{s2} represents the heat stored in the lubrication system.

The lubrication system oil supply pressure balance equation is as follows:

$$P_1A_1 - P_2A_2 + \sum F_{other} = \rho q(v_1 - v_2) \tag{24}$$

where P_1 and P_2 represent the forces generated by the pressure at the inlet and outlet, respectively, of the oil pump and where $\rho q(v_1 - v_2)$ is the inertial force of the lubricating oil.

The pressure balance equation of the lubrication system oil supply pipeline is as follows:

$$P_2 + \frac{1}{2}\rho v_2^2 + \rho g h_2 = P_3 + \frac{1}{2}\rho v_3^2 + \rho g h_3 + \Delta P_{f1} + \Delta P_{j1} \tag{25}$$

where ΔP_{f1} represents the head loss, which can be calculated according to the Darcy–Weisbach formula, and where ΔP_{j1} is the local resistance loss, which is usually calculated on the basis of the local resistance coefficient ξ.

The lubrication system oil return pressure balance equation is

$$P_4A_4 - P_5A_5 + \sum F_{other} = \rho q(v_4 - v_5) \tag{26}$$

The lubrication system oil return pipeline pressure balance equation is

$$P_5 + \frac{1}{2}\rho v_5^2 + \rho g h_5 = P_6 + \frac{1}{2}\rho v_6^2 + \rho g h_6 + \Delta P_{f2} + \Delta P_{j2} \tag{27}$$

The lubrication system mass balance equation is.

$$\frac{\vartheta \rho}{\vartheta t} + \nabla(\rho \vec{v}) - \dot{m}_l = 0 \tag{28}$$

where $\frac{\vartheta \rho}{\vartheta t}$ denotes the rate of change in the density of the lubricating oil over time; $\nabla(\rho \vec{v})$ denotes the divergence of the lubricating oil mass flux, which reflects the flow and distribution of the lubricating oil mass in space; and $\dot{m}_l$ represents the lost lubricating oil mass.

The lubrication system flow distribution balance equation is

$$q_{in} = \sum_{i=1}^{n} q_{out,i} + q_l + q_s \tag{29}$$

where $q_{out,i}$ is the flow rate of the i-th lubrication point.

The problem of solving the lubrication oil system model is similar to that of the fuel system model, with the addition of simulation calculations for thermal characteristics.

Joint Solution Scheme

As shown by the modeling introduction of the above subsystems, the modeling methods and solution calculation schemes of each discipline are different. If all the variables to be solved for and the simulation are coupled directly for the overall solution, the technical problem of difficult unified solution will be encountered, which will be manifested in two main aspects: First, the solvers required for the variables and equations to be solved by each discipline are different, and choosing or designing a general solver that can efficiently solve all problems is difficult. Second, the time steps required for the simulation solution differ substantially among specialties. For example, the hydraulic fluid equation solutions required by the fuel system and lubrication system models generally require microsecond-level time scales, while the simulation steps of the engine, control and other systems are generally at the millisecond level or above, and the differences between them are very large. If a microsecond-level simulation step is used directly for the whole system model simulation calculation, it will lead to a significant increase in the simulation calculational burden, and the model solution time will be very long; if all the models are solved with millisecond-level simulation steps, the differences between the calculation results for the hydraulic fluid simulation model and the correct values will be large, and the simulation error will be difficult to control. Therefore, owing to the differences between multiple system models, adopting a unified solution scheme for simulation calculations for the enhanced model of the whole aero engine system is difficult.

To solve the Co-simulation calculation problem for multiple system models, this paper adopts a step by-step solution scheme. The strategy is to decompose the originally coupled high-dimensional equations to be solved into multiple subsystem models according to the different specialties of each subsystem, solve them separately via the solvers required by each specialization, and solve the equations between each subsystem. Once all the solutions are obtained, the entire co-simulation model can be calculated. In this process, the correctness of the entire solution simulation calculation result is ensured because all the solution equations are satisfied; in addition, through the decomposition and separation of the solutions of the equations to be solved, dimension reduction is achieved, which can reduce the difficulty and calculational burden of the overall solution and improve the efficiency of the simulation calculation. Moreover, to address the problem of inconsistent time scales of simulation solutions of different subsystem models, the method of interpolation with a time base is adopted to align the times, thus ensuring the time consistency of the calculation results.

Application Cases of the System Enhancement Model

There is performance coupling between the subsystems of the aero engine, such as transient response performance coupling between the engine, the fuel system and control system, aerodynamic and thermal performance coupling between the engine and the air system, thermal coupling between the fuel system and the lubrication system. These performance coupling characteristics have important effects on the overall operating efficiency and safety of the engine, but the performance coupling results between the systems are difficult to calculate through single-system model simulation. The enhanced model of the whole aero engine system is used to build the model according to the physical coupling mechanism between the systems, which can address the coupling effect between the systems and effectively reflect the overall response result of the system. Here, taking the transient response coupling between the engine, the fuel system, and the control system and the thermal coupling between the fuel and lubrication systems as examples, the benefits of the whole system enhancement model in the system scheme design and verification process are introduced.

Engine–Fuel–Control System Transient Performance Verification

The transient response characteristics of an aero engine have a crucial effect on the safety and service life of the engine and are influenced by multiple disciplines, such as aerodynamics, thermodynamics, mechanical hydraulics, electronics, and control; the associated quantities generally need to be adjusted and calibrated through experimental testing. In this process, the control parameters need to be adjusted not only with changes in the working point but also at different working points within the engine working envelope. Moreover, the transient test process poses a greater risk for newly developed engines, and coverage testing of each working point and state point requires repeated debugging and verification, thus leading to high costs and long cycles.

To reduce the risk and cost of the test process, researchers use digital simulation methods to obtain preliminary design results for control parameters. However, owing to the many factors that affect the dynamic response of the engine and the involvement of multiple disciplines, efficiently and effectively integrating the models of different disciplines is difficult. Generally, simplified methods are used, or only the engine performance model and control model are integrated for analysis and design, which results in large deviations of the parameters designed in the early stage, which makes providing guidance and help for actual testing difficult. In addition, various fault problems and fault propagation effects and performance characteristics in the system are difficult to effectively reflect in the simplified model, and analyzing the system fault-in-loop response characteristics is impossible, which is not conducive to providing feature data for fault diagnosis algorithms. The actual fault injection test has a high risk, and the accumulation of fault data is small, which is not conducive to the design of fault online detection and fault-tolerant control schemes.

Through the concept of the aero engine whole system enhancement model proposed above, a multisystem co-simulation model of engine performance, a fuel system, a control system, and a sensor system is established. The establishment of a professional mechanism model and parameters for each system can effectively reflect the effect of the dynamic performance of each part on the whole system response characteristics, which is conducive to the adjustment and optimization of control parameters and provides a reference for actual test parameter debugging.

Figure 9 is based on a transient performance characteristic simulation conducted using the established co-simulation model. The simulation results show that the control parameters designed without considering the fuel system or sensor system response characteristics deviate from the expected effects after introducing the whole machine system enhancement model for co-simulation. This requires the redesign and adjustment of the control parameters, thus causing much of the previous work to be redone, affecting R&D efficiency and increasing costs. By introducing a multidisciplinary coupling model, the dynamic characteristics between the aero engine and the fuel system can be simulated during the computer simulation stage to reflect the phenomena and problems that exist in the actual physical system and provide effective support for control system design and debugging (Fig. 9).

Fuel and Lubricating Oil System Thermal Coupling Simulation

The fuel system and the lubricating oil system are not independent systems, and there are thermal interactions among the lubricating oil system, the fuel system and the air

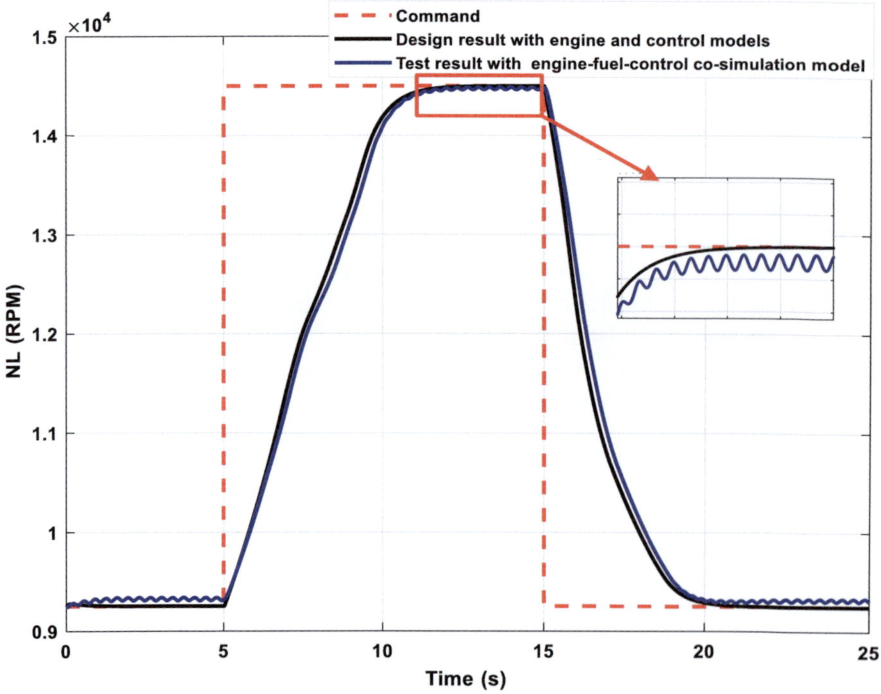

Fig. 9 Simulation curve of the aero engine transient response process

system [15, 16], where the fuel system and the lubricating oil system interact thermally through the fuel/lubricating oil heat exchanger. Low-pressure fuel enters the fuel/lubricating oil heat exchanger to cool the lubricating oil, and after the lubricating oil is heated, the fuel enters the high-pressure pump to increase the pressure and is supplied to the engine for combustion through a metering device. The air system and the lubricating oil system interact thermally at the bearing chamber, where the air system maintains the bearing chamber pressure through the sealing device while hot air leaks into the lubricating oil system, thus resulting in two-phase flow heat transfer. The coupling relationships among the fuel system, the lubricating oil system, and the air system are shown in Fig. 10.

As the engine Mach number gradually increases, the heat generated by the lubricating oil system also increases, and the heat exchange between the fuel and lubricating oil becomes more prominent in its effect on engine performance. However, in the current engineering design process, owing to limitations such as the development unit and software interface, the fuel and lubricating oil systems are currently modeled separately, and only their own pressure and flow requirements are usually considered, without considering heat exchange between systems or using empirical data for simple heat exchange estimation. This design method not only does not conform to the actual working conditions of the engine but also, in extreme cases, may cause overheating of the fuel system or the lubricating oil system because of mismatched

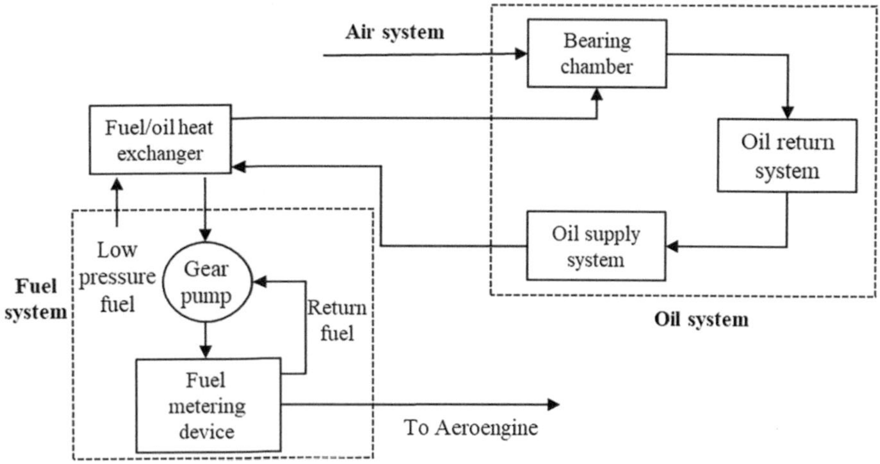

Fig. 10 Relationship between the fuel/lubricating oil system and the air system

thermal coupling between systems, thereby affecting the normal operation of the entire engine.

To verify the effect of heat interactions between fuel and lubricating oil on the system, a model of the interactions between the fuel system, the lubricating oil system, and the air system was established on the basis of a certain type of turbofan engine, and a co-simulation was carried out for heat exchange between the systems, which demonstrated the effects of heat interactions on the lubricating oil system and the fuel system.

After the design of the fuel system and the lubricating oil system of a certain type of engine was completed, combined with the changes in engine operating conditions, and after integrating the engine performance, control, fuel, lubricating oil, air and other systems for integrated simulation, we found that the increase in the fuel system would occur the overheating problems at various high-altitude operating points, as shown in Fig. 11. This problem was not found in the separate design simulation process of the fuel system. However, after integrated simulation with the engine and other system models, owing to the thermal coupling problem between the fuel system and the lubricating oil system, the usage requirements could not be met at all operating points of the whole machine, and joint design optimization work needed to be carried out. Therefore, conducting multidisciplinary and multisystem co-simulations, conducting comprehensive analyses of the performance of various related systems of the engine, and predicting the performance of the engine in actual flight can help reduce the risk and cost of actual testing, shorten the R&D cycle, and improve the engine performance.

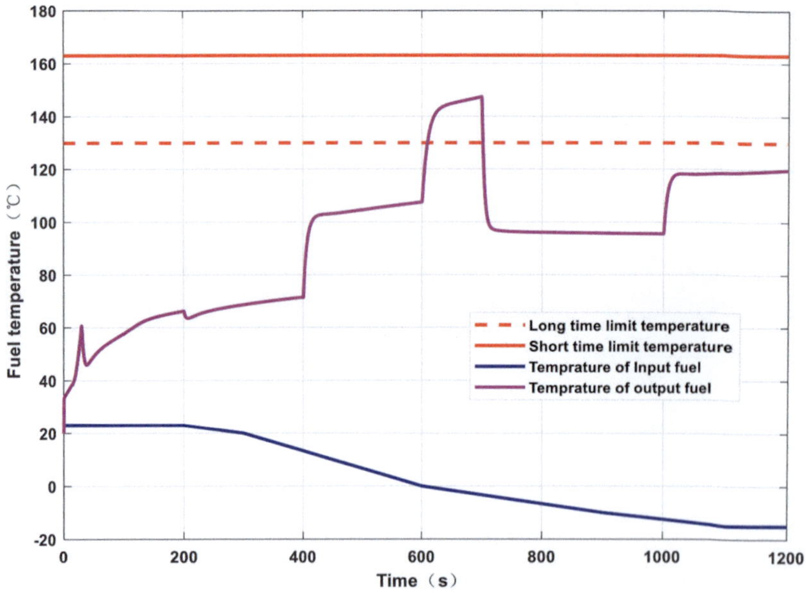

(a) Fuel system temperature curve

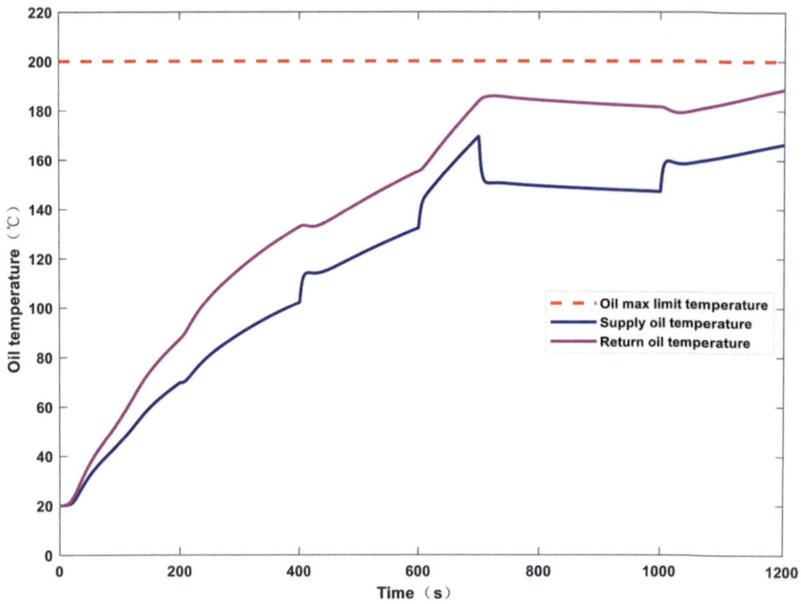

(b) Lubricating oil system temperature curve

Fig. 11 Co-simulation results for fuel/lubricating oil

Conclusions

This article proposes the concept of an aero engine whole machine system enhancement model and a multidisciplinary coupling degree evaluation standard on the basis of the coupling mechanism between multiple systems of the aero engine. It defines the signal coupling method of the aero engine multisystem co-simulation model from the professional perspectives of aerodynamic thermodynamics, mechanical hydraulics, electronics and electricity, thus providing a reference for effectively reflecting the interconnection relationships between various components and systems of the aero engine. In addition, taking the dynamic characteristic coupling problem among the engine performance, the fuel system, and the control system and the thermal coupling problem of the fuel and lubricating oil as examples, this article introduces the modeling process and solution strategies for the aero engine whole machine system enhancement model and explains the value of the system enhancement model in scheme design and analysis through simulation results.

References

1. Jianguo C (2018) Research status, challenges and prospects of aero engine simulation technology. Propulsion Technology 39(05):961–970
2. Nichols L, Chamis C (1991) Numerical propulsion system simulation—an interdisciplinary approach, NASA
3. Evans A et al (1998) Numerical propulsion system simulation's national cycle Program.In: The 34th AIAA joint propulsion conference and exhibit, AIAA: Cleveland, OH, U.S.A
4. Lytle J (1999) The numerical propulsion system simulation: a multidisciplinary design system for aerospace vehicles
5. Csank J et al (2019) Electrical power system sizing within the numerical propulsion system simulation. In: AIAA propulsion and energy 2019 Forum, AIAA: Indianapolis, IN, U.S.A. pp 1–8
6. Chapman JW et al (2014) Propulsion system simulation using the toolbox for the modeling and analysis of thermodynamic systems (T MATS). In the 50th AIAA joint propulsion conference, AIAA: Cleveland, OH, U.S.A. pp 1–14
7. Ivanov M, Nigmatullin R Interconnected Multi-level Design of Gas Turbine Elements, In: 41st aerospace sciences meeting and exhibit
8. Larsson T, Homsi P (2014) VIVACE—value improvement through a virtual aeronautical collaborative enterprise
9. Bala A. et al (2007) PROOSIS—a collaborative venture for gas turbine performance simulation using an object oriented programming schema. In: Proceedings of the 18th ISABE conference. Beijing, China
10. Alexiou A (2014) Introduction to gas turbine modelling with PROOSIS—2nd edn. Empresarios Agrupados Internacional S.A. (EAI)
11. Daxiang L, Jie J, Denghuan L (2022) The status and role of numerical simulation technology in aerospace power development. J Aerosp Power 37(10):2017–2024
12. Cao Zhipeng et al (2008) Application of CANSS1.1 system. In: Proceedings of the first academic exchange conference on numerical simulation and digital design of aero engines by the Chinese society of aeronautics and astronautics, Beijing, pp 1–4
13. Modelisar (2010) Functional mock-up interface for model exchange.
14. Spang Iii AH, Brown H (1999) Control of jet engines. Control Eng Pract 7(9):1043–1059

15. Sun H, Guo Y, Zhao W (2019) Co-simulation of aircraft engine fuel and oil system. In: The 10th international conference on mechanical and aerospace engineering (ICMAE), IEEE
16. Yang S et al (2023) Transient simulation on flow and heat transfer characteristics of aeroengine fuel-oil thermal management system. In: Asia-Pacific international symposium on aerospace technology proceedings, Hainan, China

Research on a Construction Method for Digital Thread Based on a Central Semantic Model

Abstract The digital research and development of aero engines requires breaking down the information barriers between the lifecycle development processes, constructing authoritative data sources and unified communication platforms for the product, and linking information from all stages of engine development to achieve digital decision-making. This paper takes an aero engine as the research object and proposes a digital thread construction method that is based on a central semantic model. It develops a unified metadata definition standard that is based on the concepts, knowledge, behavior, intentions, logic, and relationships of aero engines, as well as an engine feature semantic standard component library for different scenarios. Through the method of component assembly, a semantic mapping model of models at various stages and disciplines is established, thus forming an authoritative data source on the basis of the central semantic model. By using an automated mapping universal interface for heterogeneous models, bidirectional seamless integration with the semantic model is achieved, thus maintaining the consistency and traceability of information and supporting the automated construction and complete expression of digital prototypes. Finally, with the specific application scenario of CAD structural data penetration as an example, application research is carried out, which provides a theoretical and technical reference for the promotion of digital transformation in industry.

Keywords Digital thread · Authoritative data source · Semantic model · Metadata · Aero engine

Introduction

With the increasing demand for digital equipment development, the full lifecycle development of aero engines is gradually moving toward systematization and integration. The development process of aero engines has begun to consider the interactions and integration of data and data, data and models, and models and models at different development stages. With the advancement of system engineering methodology, the

Q. Xu et al., *Digital Intelligent Aeroengine*,
https://doi.org/10.1007/978-981-95-0942-3_4

use of digital technology throughout the lifecycle of aero engines is changing the way aero engine products are designed, manufactured, integrated, verified, and maintained. Currently, model-based system engineering (MBSE) is being applied in the research and development of world-renowned aero engine manufacturers. The unified collaborative platform and data management architecture of digital thread have gradually increased in scale. The innovation of existing processes and model libraries to form multidisciplinary integrated virtual verification capabilities has achieved initial results, and the importance of data fusion and intelligent decision-making has gained widespread consensus.

Digital threads address the collaboration and consistency of information, data, and models of various participants and specialties in complex system engineering to ensure a unified authoritative data source and data-driven automation throughout the lifecycle. The construction of a digital thread achieves at least the following benefits: (1) the synchronization and iterative update times of data in multiple systems and heterogeneous formats are greatly reduced, (2) the error and uncertainty caused by collaboration are reduced or even eliminated, and (3) the completeness of equipment status expression is ensured to support mapping. The digital thread extends the model-based definitions to all activities throughout the product lifecycle, uses models to drive various business activities throughout the product lifecycle, ensures data consistency and reusability, and achieves model continuity. Moreover, only with a digital thread as the digital foundation can a unified engine digital model be formed throughout the lifecycle, thus supporting the construction and penetration of a digital prototype model of the equipment. Research in China and other countries on the architecture definition and implementation plan for the engine digital thread has been systematically carried out, which has included the digital thread architecture, which is based on the semantic integration of Pratt & Whitney (P&W) [1], and the design-simulation integrated architecture of Safran [2]. Additional work has been carried out to implement complex system modeling work for engines with different MBSE methodologies as the goal, and a unified modeling method and construction process that are based on engine professional knowledge are lacking; however, through the integration of system models and various professional models, models can be seen but not used, and their connections can only be matched manually by industry experts, which does not support automatic reconstruction, thus making obtaining obvious benefits such as reduced iteration time difficult. The development of a digital thread architecture that runs through the entire lifecycle of the engine to support the construction of a digital prototype of the engine is urgently needed.

This paper takes an aero engine as the object and proposes a digital thread construction method that is based on a central semantic model. It develops a unified metadata definition standard and a standardized component library of engine feature semantics for various scenarios on the basis of the concepts, knowledge, behavior, intentions, logic, and relationships of aero engines. Through the method of assembling the component library, a semantic mapping model of models at various stages and disciplines is established, thus forming an authoritative data source on the basis of the central semantic model. Using an automated mapping universal interface for various professional models, it achieves bidirectional seamless integration with the semantic model, thus maintaining the consistency and traceability of information throughout the lifecycle and supporting the automated construction and complete expression of digital prototypes.

Technical Approaches from China and Other Countries

The digital thread introduces a data communication architecture where resources (such as sensor outputs, computational tools, methods, and processes) can be shared, thus linking information from all stages of the product lifecycle (such as early concepts, design, manufacturing, maintenance, and retirement) to achieve real-time and long-term decision-making. It can also be considered the product's main or "authoritative" data and communication platform [3].

Tracing back to the source, the concepts of digital thread and authoritative data sources come from the further expansion of the internet semantic web, ontology modeling, and knowledge graph technology. The semantic web makes the entire internet a universal information exchange medium by adding semantics (metadata) that can be understood by computers. Semantic web technology is based on a set of standard languages, models, and tools that are suitable for knowledge representation, including extensible markup language (XML), the resource description framework (RDF), the web ontology language (OWL), and the RDF query language and protocol (SPARQL). [4] As shown in Fig. 1, by describing the instances of the ontology (specific model instances) and the metamodels that characterize the information and relationships of the instances, the current internet or specific application scenarios are reconstructed to enable computers to obtain meaning from information to achieve automated information processing.

The core underlying layer of the data is the syntax layer, which is composed mainly of XML and related specifications. This layer can highlight the description of the structure while describing the data content and freely defines the tag names and element hierarchy in XML. Above the syntax layer is the resource description framework, which includes mainly RDF and related specifications. It can be regarded as an XML description with a specific format in which a triplet composed of a subject, predicate or attribute, and object or attribute is used to describe the metadata of the resource. On the basis of the RDF, RDF schema and the OWL are defined to express complex concepts, rules, and interactive relationships, which can support computers in reasoning in knowledge bases and ontologies.

For the authoritative data source and digital thread method of digital engineering, the Naval Air Systems Command (NAVAIR) of the United States proposed the digital engineering integration and interoperability framework (DEFII) [5], as shown in Fig. 2, which is based on the expansion of semantic web technology to solve the problems of multisource heterogeneous data and full lifecycle activity integration and model interoperability faced in digital engineering, thus aligning engineering data and semantic ontology and making data processing independent of specific tools. The bottom layer of the architecture is the data layer aligned with the ontology, which

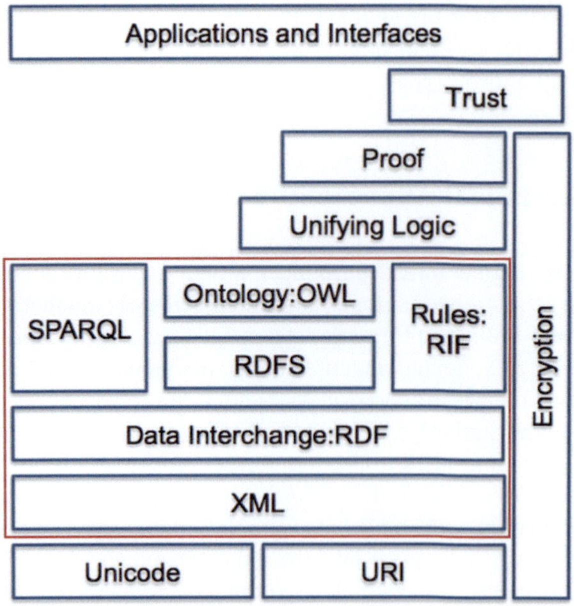

Fig. 1 Semantic web technology architecture [4]

mainly defines the domain knowledge ontology and then aligns the integrated data from various tools to the ontology. The ontology knowledge base uses a controlled "dictionary" defined by ontology class data to describe actual engineering data as instances of specified classes. Reasoning enables automatic reasoning functions, establishes relationships between classes through automatic reasoning, and does not need to explicitly express all knowledge. The ability to infer knowledge on the basis of heterogeneous data storage opens up the possibility of discovering how design elements from different tools are related to each other. The interface layer establishes, updates, views, and analyzes the interface between the authoritative data in the system and external data sources. The framework defines three types of interfaces: direct, mapping, and designated model interfaces. The direct interface refers to the alignment and processing of semantic data and direct access using semantic operation tools such as SPARQL; the mapping interface specifically reads the MBSE system model and maps it to semantic data; and the designated model interface provides an interface with external multidisciplinary and heterogeneous tools and applications so that external tools can obtain relevant system semantic data.

In the specific industry of aero engines, Pratt & Whitney uses digital engineering as a way to connect different tools and optimize automation by performing cross-domain integration through the digital thread and forming the architecture of the digital engine platform [1]. As shown in Fig. 3, its core strategy is to use semantic data integration as a data source and use a unified data integration platform to integrate all the models and data from different systems.

For specific unified heterogeneous data expression integration, the German Aerospace Center (DLR) Development Center data model CPACS [6], as shown in

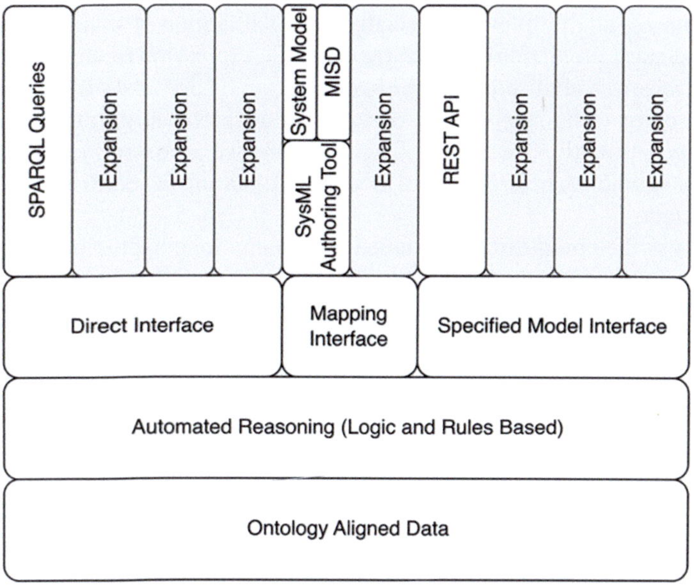

Fig. 2 Digital engineering integration and interoperability architecture [5]

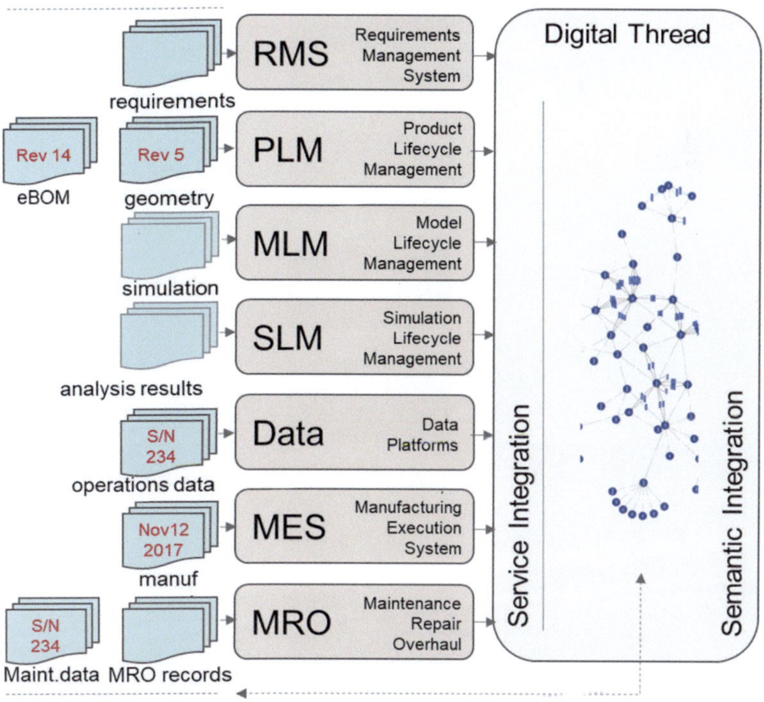

Fig. 3 Pratt & Whitney's digital thread integration architecture [1]

Fig. 4, combines all disciplines (especially the collaboration of structural design and multidisciplinary simulation) through the semantic expression of standardized XML data and exchanges all discipline information through XML data files, with general parameterization technology used to describe the aircraft configuration scheme with high fidelity. Currently, this project is aimed at the overall design of the engine and lacks the integration of heterogeneous models and data and the control of the engine's technical status.

To address the complexity of detailed design and manufacturing processes, the National Institute of Standards and Technology (NIST) in the United States extended the product manufacturing information (PMI) data model to the ISO 10303 standard range, developed semantic-based PMI expressions, and considered readability by engineers and the operability of computers so that engineering software developers can automatically read PMI when CAD/CAM/CMM functions are integrated [7], as shown in Fig. 5. This method extends the expression of semantic language to part-manufacturing information.

Similarly, semantic language can be used to describe and define data from other stages of a product's lifecycle. The University of Nottingham used XML to define and describe product design models, process models, and equipment models and used a unified language to connect and provide feedback among the design department,

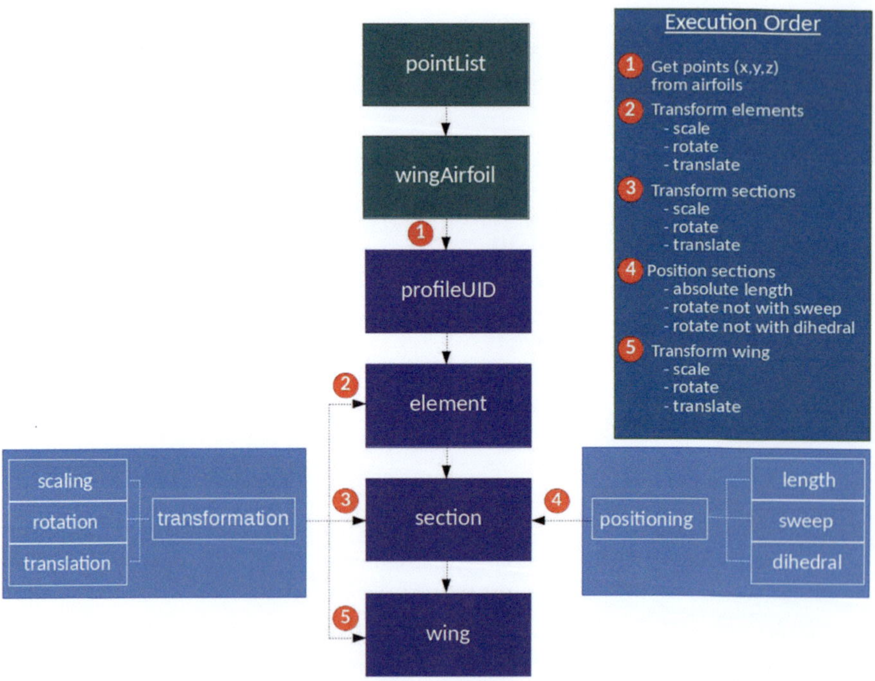

Fig. 4 Aircraft center data model (CPACS) [6]

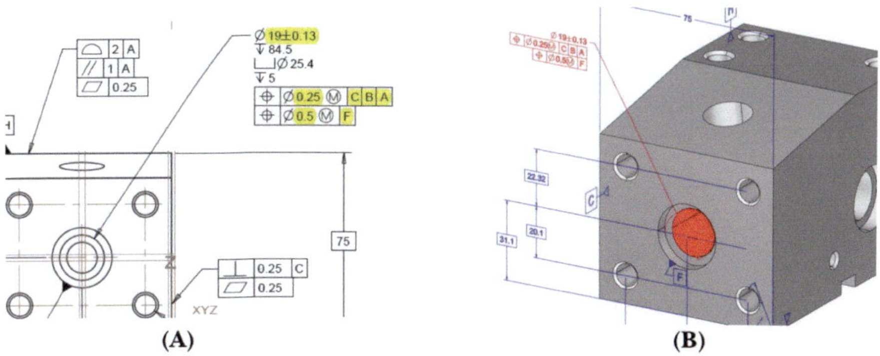


```
<?xml version ="1.0" encoding="UTF-8"?>
<ModelTree3D Version = "1.0">
  <Node Name="DIM">
    <Property Name="cadEntityId" Value="1339"/>
    <Property Name="dimensionText" Value="{0: n }{1:@D}{2: # 0.13}"/>
    <Property Name="dimensionType" Value="DIMTYPE_DIAMETER"/>
    <Property Name="displayedValue" Value="19.000"/>
    <Property Name="gtl_gtolType" Value="GTOLTYPE_POSITION"/>
    <Property Name="gtl_isComposite" Value="false"/>
    <Property Name="gtl_isOverallTolerance" Value="true"/>
    <Property Name="gtl_isPerUnitTolerance" Value="false"/>
    <Property Name="gtl_materialCondition" Value="GTOLMATCOND_MMC"/>
    <Property Name="gtl_overallToleranceValue" Value="0.250000"/>
    <Property Name="gtl_primaryBasicDatumName" Value="C"/>
    <Property Name="gtl_secondaryBasicDatumName" Value="B"/>
    <Property Name="gtl_showAllAroundSymbol" Value="false"/>
    <Property Name="gtl_showDiameterSymbol" Value="true"/>
```

(C)

Fig. 5 XML Expression of PMI [7]

manufacturing department, and factory equipment department to create an environment for manufacturability assessment in the early stages of design [8] (see Fig. 6). Ohio University and the Delphi Motor Company developed the APPS system, which uses XML files to express the manufacturing feature information of the process model for the integration of the CAD, CAPP, and CAE environments [9]. Research has also been carried out in China on local application scenarios of the equipment manufacturing process. Northwestern Polytechnical University generated XML files from the extracted process MBD model information for integration and invocation with the CAE machining environment [10]. Nanjing University of Science and Technology developed the PDAS system, which describes the product assembly model library with XML and realizes the reusability of component models and product assembly models by searching for and modifying the component models and product assembly models in the library [11].

In terms of the tool platform integrated with the digital thread, the Intercax company developed the software platform Syndeia, which establishes a platform that is connected with other engineering models (such as TeamCenter, Windchill, SQL, CAD, and CAE) modeled with the SysML language [12], as shown in Fig. 7, using system modeling language to integrate and link information between tools and to extract and analyze specific model information or single-stage data information effectively.

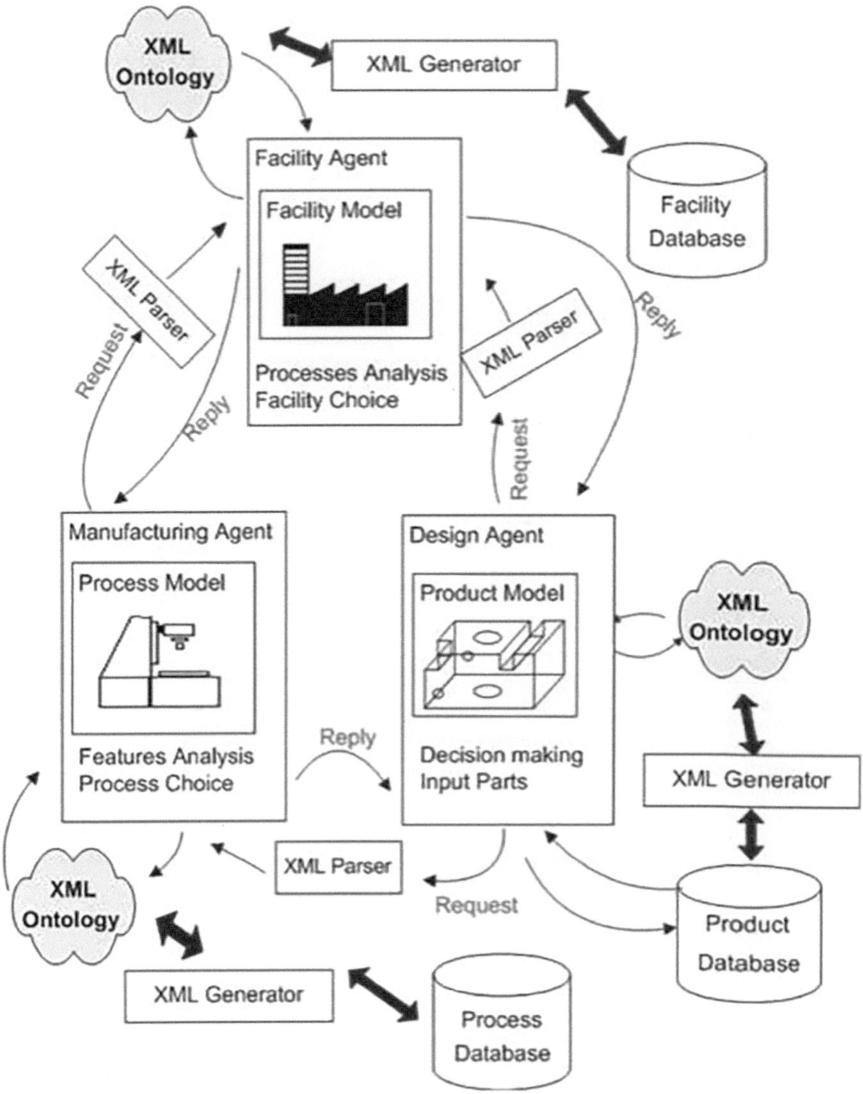

Fig. 6 Manufacturability assessment architecture [8]

The National Institute of Standards and Technology (NIST) in the United States built a data integration platform [13] that converts QIF files (based on XML) and STEP files (based on EXPRESS) into SWRL files on the basis of semantic web rules, establishes mapping relationships, integrates knowledge bases, and enables designers and inspectors to use semantic language to query the knowledge base. With the richer logic and reasoning capabilities of the SWRL language, the platform realizes information extraction and web-based queries, but when different model

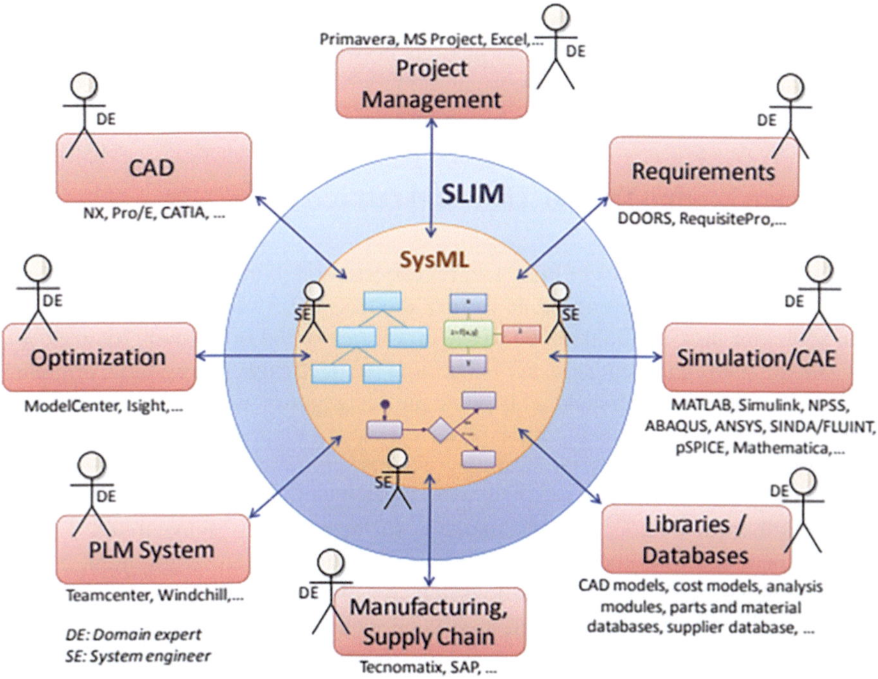

Fig. 7 Syndeia's heterogeneous data integration platform [12]

data are converted into SWRL knowledge graphs, mapping relationships need to be established, which can be extremely complex if extended to the entire product lifecycle.

To implement a digital thread, consideration should also be given to integrating a unified digital base for multiple systems. Huawei's iDME platform has been used to develop dynamic data access technology that supports multiple protocols and a digital thread model-driven engine. Tsinghua University has built an industrial multisource heterogeneous big data integrated management analysis engine, IGinX [14], which provides access and queries for different information. The core challenge of current systematic exploration of heterogeneous data integration engines and dynamic data access on the basis of big data platforms is addressing the complexity of actual scenarios and domains.

In summary, a technical approach that is centered on semantic models is applied to all stages of a product's lifecycle, whether in model and process expression or data link integration, and it is already the most likely path for a digital thread. However, recent research and tool development have focused mostly on single-point, local applications and have not yet formed a complete digital link. There is also a lack of effective solutions for the complexity of specific objects. With the continuous development of model-based systems engineering (MBSE) methods, there is an urgent need to develop a digital thread technical architecture that spans the entire product lifecycle to realize model information interactions, including interactions between disciplinary knowledge, interactions between heterogeneous data models, and information communication between models.

Aerospace Engine Digital Thread Architecture

Given the complex multidisciplinary heterogeneous models and data within the entire lifecycle of an aerospace engine, to ensure the traceability and consistency of the data, upstream data can be updated to the downstream end in a timely manner, and downstream data can be updated and fed back to the upstream end in a timely manner, while each stage can understand and update the design intent when receiving external data. Currently, within the lifecycle of an aerospace engine, there is still a large amount of data from intent and relationships that are not controlled, uncertainty is continuously introduced, and there are even more data for which relationships need to be established, without suitable methods for establishing associations. For the establishment of requirement models, the interactions between requirement models and discipline models, or the interactions between discipline models or the integrated use of engineering software applications at each stage, large amounts of interpretation and translation by engineering personnel are required, and the data cannot be read directly by computers to achieve computer operability of the digital thread of the entire lifecycle of an aerospace engine.

The semantic expression languages that are currently recognizable by computers are mostly based on XML. XML is an extensible and highly flexible markup language that can be used to describe and express underlying base languages and can serve as a semantic expression of data for backend processing. Users can define their own tags according to their design needs and can use semantic language to express knowledge and data in various fields in detail, thus providing a format method for structuring and sharing data. Moreover, XML has good software and WEB compatibility, can build data knowledge bases well, and can be well recognized and processed by computers. Various language formats that are involved in semantic web technologies, such as RDF, RDFS, and OWL, are all based on XML and can be viewed as XML instances that have more specific syntax structures to better describe resources and ontology relationships.

System models mainly describe the composition and association of products from a global perspective through modeling languages such as SysML in the form of semantic modeling and can also provide management media for models to drive or guide the integration and optimization of models from a management perspective. Products such as aircraft engines with high complexity and strong coupling support the analysis, specification, design, verification, and confirmation of complex systems (including hardware, software, information, processes, personnel, and facilities) and can express the structural composition, association relationships, and physical performance attributes of the system. The formalized modeling language SysML can be understood as supporting human–computer interaction in system modeling by enabling the roles, relationships, dependencies, and interfaces between elements to be defined and understood by developers. In system modeling software, data in SysML format are generally also converted into XML format for storage and processing, which provides us with the possibility of XML-centered data.

On the basis of the analysis of semantic expression in various languages, this paper proposes a method for constructing a central semantic authoritative data source for aircraft engines on the basis of the XML semantic model, as shown in Fig. 8.

For several application scenarios in various life cycles of aero engines, we establish multisource, cross-domain metadata standards and specifications on the basis of domain knowledge and build a semantically correct, independent and unambiguous metadata definition set of specific concepts through identifiers, representations, allowable values and other attribute description units. Typically, the construction of the metadata definition set requires manual development by industry experts. There are also technical routes that aim to intelligently extract metadata definitions on the basis of technical reports and industry standards, but the current design system data volume has difficulty meeting the requirements of digital prototypes in terms of granularity and explicit expression of relationships, and the uncertainty of artificial intelligence is also a problem. On the basis of the constructed metadata, the core issue is what the smallest unit of complete engine expression is. Taking the structural model as an example, the intermediate-format STEP model exported from the NX structural model includes all the information about points, lines, surfaces, and bodies, whereas designers are concerned with structural feature information such as flanges, holes, shafts, and bosses. By analogy to other heterogeneous models, including MBSE system models, model features are the smallest meaningful units, and their attribute definitions are the smallest level of concern for R&D. Therefore, we construct an aero engine lifecycle feature semantic description component library (XML model) through feature recognition and extraction and assemble a central semantic model that can be mapped with various models on the basis of the feature component library. The core requirement is that search and navigation are unique in the global range, and repeated expressions should be referenced. By constructing an engine's central semantic authoritative data source as the main body of technical status control, unified language descriptions of demand models, architecture models, structural models, simulation models, process models, implementation models and maintenance models can be realized throughout the entire lifecycle of aero-engines and the authoritative data source can be continuously instantiated during the entire lifecycle, thus realizing the synchronous update of data models and ensuring the consistency of data throughout the product lifecycle.

On the basis of the construction of the authoritative data source, this paper proposes a digital thread platform architecture that meets the requirements, as shown in Fig. 9, which is divided into a basic layer, a heterogeneous data layer, a digital thread layer, and an application layer. The basic layer consists of the basic hardware infrastructure and system standards for the digital transformation of equipment and the construction of digital prototypes. The heterogeneous data layer, which is based on a common IT basic software platform, continues the digital systems at different stages of the

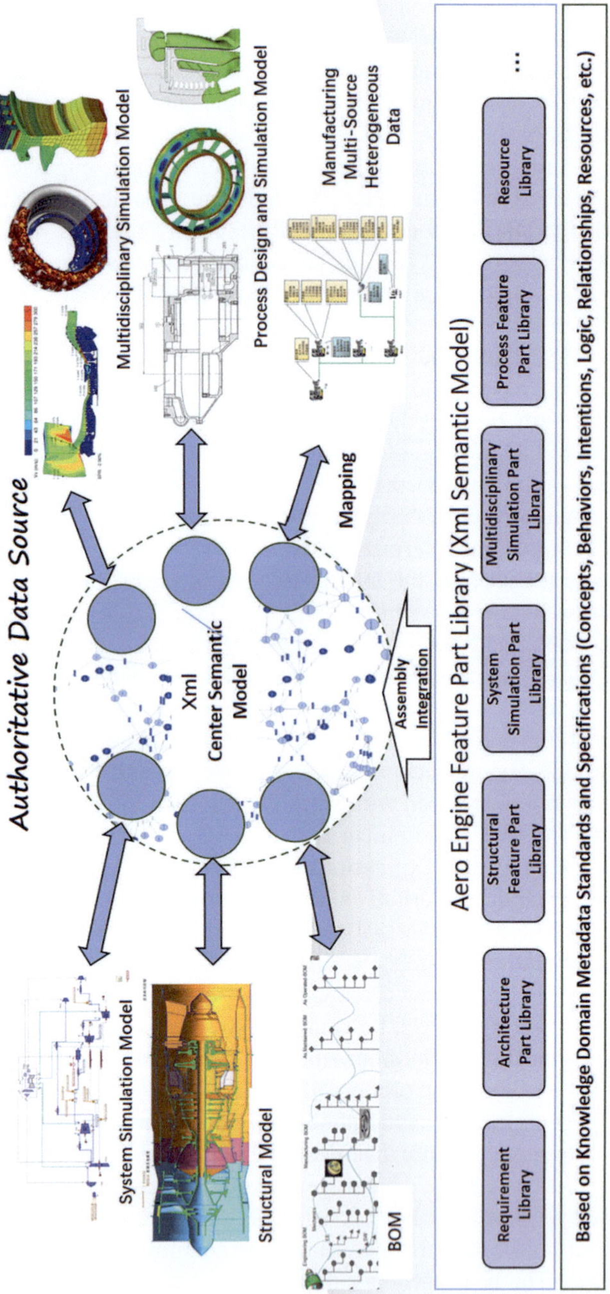

Fig. 8 Semantic model-based authoritative data sources

lifecycle contained in the current R&D system, including multisource heterogeneous data integrated in different systems.

The heterogeneous data layer also needs to build a unified data integration engine base to achieve rapid data query and association calculations. The current commonly used technical routes all have problems. For two systems to carry out data interaction and communication, users must perform processing with complexity $C(n_2)$ every time a data association calculation is involved; the calculation is not only of high complexity but also of poor scalability. The scheme for integrating data into an intermediate system that is based on the data middle platform requires regular data transfer. As the amount of data increases, its utility gradually decreases, thus increasing system complexity; it cannot guarantee real-time synchronization of data between the middle platform and other systems, and the implementation of association rules is difficult.

This article proposes a unified data integration engine, as shown in Fig. 10, which can serve as efficient middleware for integrating heterogeneous multisource data, thus enabling connectivity with other data, collecting and synchronizing information in a given format, breaking through the unified management and query of heterogeneous multisource data, and further supporting data query evolution capabilities. Through joint mapping from the heterogeneous data space to the pandata space, a unified data space is realized, thus developing multimodal data query technology for SQL-like relational query language, supporting complex calculations of multimodal data through specific model integration, expressing data evolution relationships, directly integrating multisource heterogeneous data, and making data directly usable and directly associated for queries. A storage-computing layered architecture is adopted, which is deployed on the basis of the underlying heterogeneous multisource data storage and constitutes a computing processing layer that is composed of a cluster of heterogeneous multisource data management middleware. The heterogeneous multisource data integration processing engine adopts a stateless service mode, so it can be expanded on demand in real time.

The digital thread layer includes three parts: an aero engine feature component library, an authoritative data source that is based on central semantics, and an automated generation and real-time driving interface. The authoritative data source can be viewed as a slice of the current technical status information that contains all complete R&D information and data, including associated relationships and iterative updates as activities progress throughout the lifecycle.

The semantic models of various disciplines of the authoritative data source are assembled through the feature component library, and after generation, they are updated and run through all heterogeneous data as needed via automated generation and the real-time driving interface of the unified heterogeneous data integration engine. The interface is divided into three parts. The first part is the logic reasoning rule library and driving algorithm that directly processes XML data, which can extract relationships and hierarchical XML semantic expression information to search and drive assembly and changes and includes the extraction of data sources for display, heterogeneous information synchronization, data analysis decision-making, etc., during the process. In the implementation of the MBSE system modeling process, constructing an automatic mapping method and interface from the SysML model to

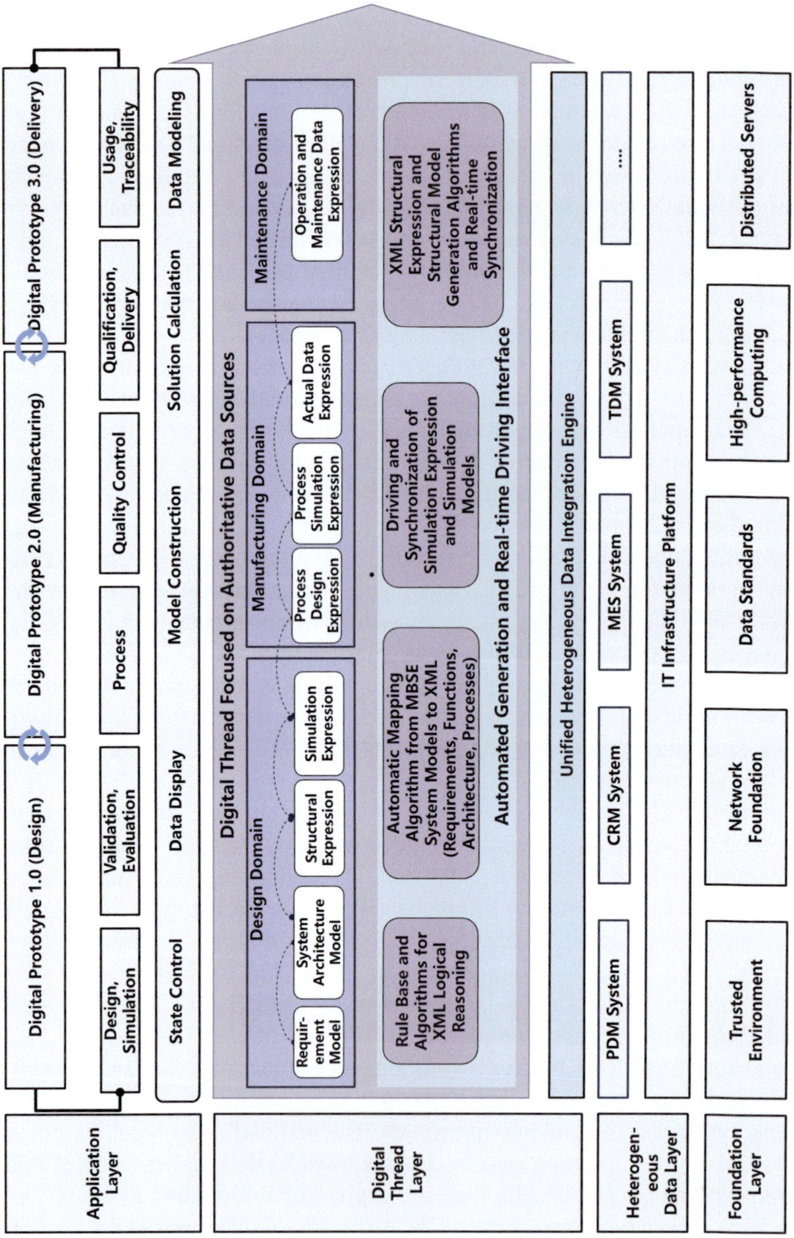

Fig. 9 Aeroengine digital thread platform architecture

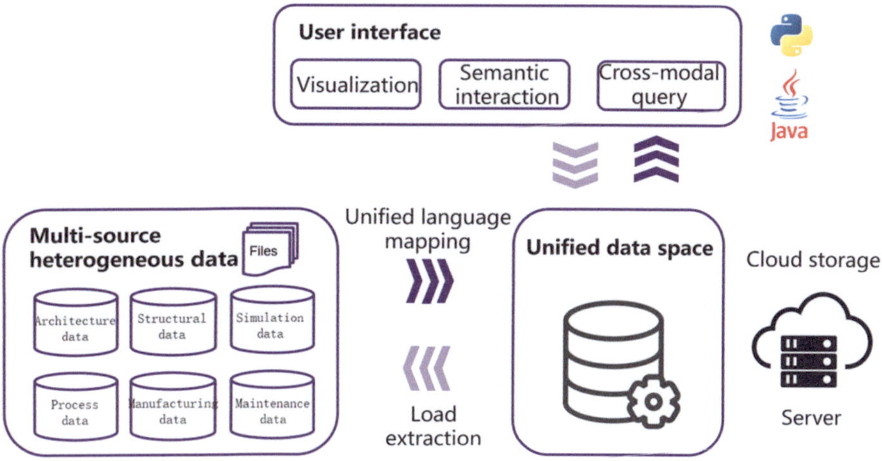

Fig. 10 Unified data engine base architecture

the XML model is necessary, which can transform the formalized model based on software into an XML expression model that meets the metadata standard specification to achieve alignment and thread integration of requirements, functions, architecture, and physical content. For specific applications (CAD, CAE, CAM, etc.), a general automated driving interface that synchronizes with the thread data is developed, which can automatically collect relevant content of the semantic expression model, implement heterogeneous model generation and update synchronization.

The application layer built on the digital thread can support data integration and visualization, data decision-making, etc.; can provide all the data and models needed for all activities of the R&D lifecycle; and can automatically complete the construction of digital prototypes. Finally, although this article proposes an ideal digital thread architecture, there are still incomplete situations caused by the complexity and uncertainty of the aero engine reaching certain levels. A feasible approach is to use the authoritative data source as a training set to support the training of artificial intelligence (machine learning, expert systems, large models, etc.), empower the intelligence of the entire architecture, further improve the automation of modeling, and reduce the difficulty of application.

Application Research

The aero engine digital thread involves a large amount of heterogeneous data and is difficult to integrate. Related work should start from typical data formats as a preliminary technical route verification.

Throughout the entire R&D process, a structural model that carries information at different lifecycle stages is the core output. Hierarchical decomposition, module connection, materials, weights, modeling features, PMI, and feature processing are all connected with the structural model. However, the model data of CAD software, which is a typical professional software, are not explicit. The related formats, such as the STEP242 models that it transforms, are complex models that describe geometric information, have very poor readability and standardization, cannot express design intentions, and do not achieve digital thread penetration of all data. Therefore, the interactive externalization of structural data is a key issue for data penetration. Constructing a feature XML semantic model that satisfies the design objectives of expressing key information and realizing real-time synchronization of the XML model and the structural model through a general script is necessary. This article reports corresponding research work for specific application scenarios of scheme design and detailed design.

Structure Expression and Modeling Based on a Central XML Semantic Model

To break through the overall integrated design platform of the digital thread, this study investigated the construction of an XML model of the two-dimensional cross-sectional structure and three-dimensional skeleton model structure of a turbofan engine and the driving of CAD automated modeling. The overall structure of the turbofan engine is described semantically in XML.

The overall structural framework of the XML description file is shown in Fig. 11. The XML overall framework includes a header file and the Engines engine file. The name (engine technical status), version, and version information content are defined in the header file. The developer, timestamp, description, and version are described in the version information content.

The engine framework mainly stores data on the overall structural components of the engine. The model structural framework divides the main body parts of the engine into structures, including fan components, intermediate casing components, compressor components, combustor components, turbine components, rear support casing components, jetpipe components, and intake cone components, and considers the interface information between each pair of components. Each component node level corresponds to the part-component structural framework information that describes the logic, dependency, and correlations between components, as well as attribute definitions, geometric features, and size parameters (Fig. 12).

The three-dimensional skeleton control model's central structured data XML semantic model ensures correlation and dependency with the preliminary data; refines the structure and reinforcement information on the basis of the two-dimensional structure model; inherits the logic, dependency, and correlation between the two-dimensional model data; maintains a unified data source; and forms a standardized database of the three-dimensional skeleton control model. The standardized database essentially links model elements with semantics to obtain an interface freedom set,

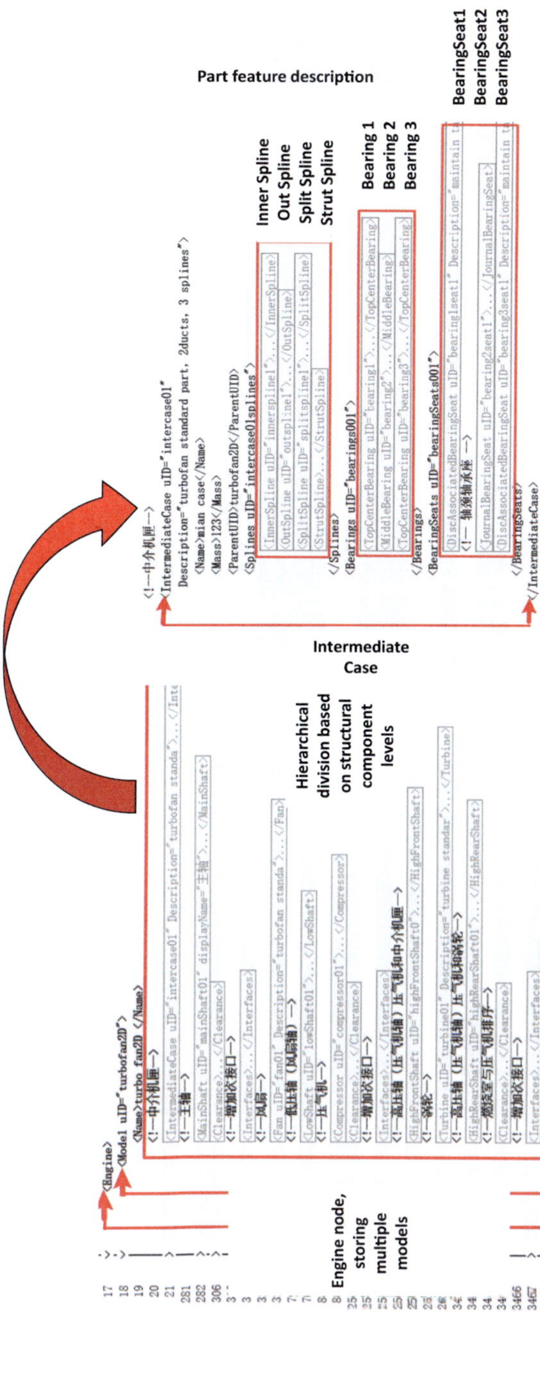

Fig. 11 Two-dimensional structural expression XML semantic model of a turbofan engine

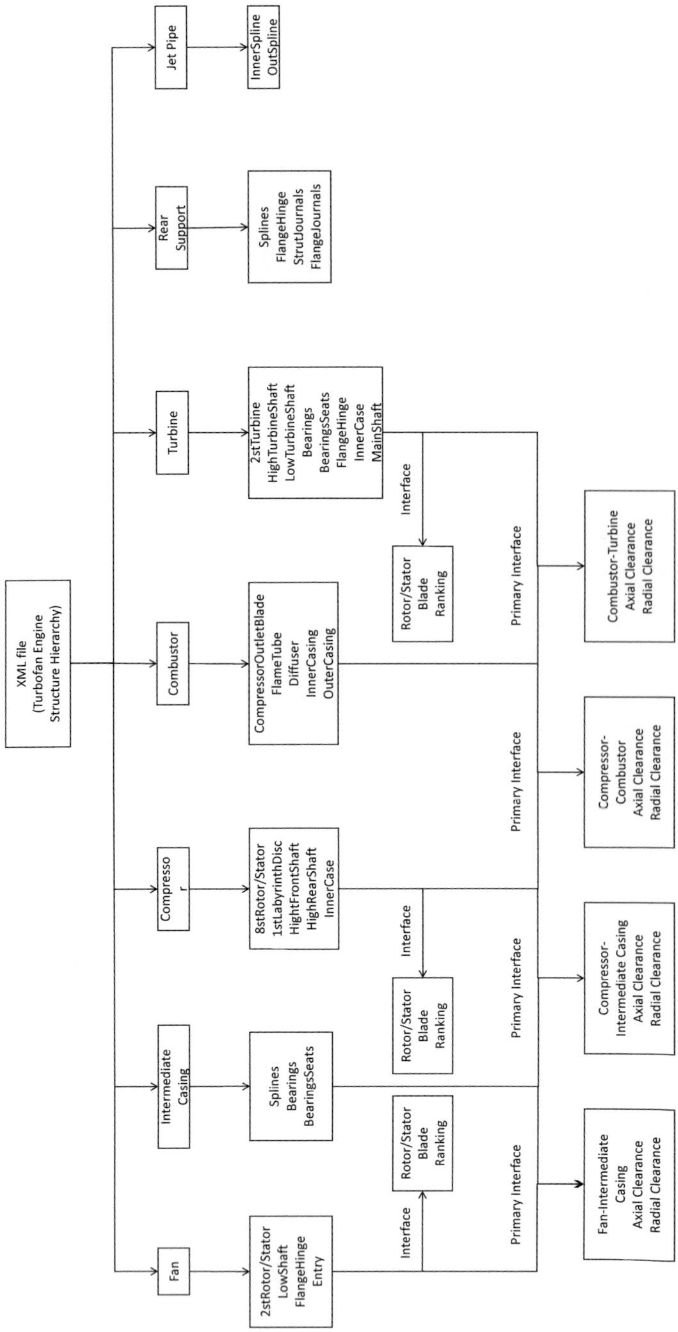

Fig. 12 Two-dimensional structural expression framework of a turbofan engine

connection parameters, and a topological structure. In the subsequent design process, only the working object is referenced without copying.

The central expression data that carry the engine structure's logical and dependency relationships cannot interact with the modeling software directly. Through the development of automated scripts, an interaction between the central structured data and the modeling software is established. The automated script uses the object-oriented C# programming language to integrate software for secondary development of NX modeling software, thus forming executable program files, realizing the mutual interaction between the central structured data XML model and NX modeling software, and passing information between them. The development of automated scripts requires the establishment of an aero engine parts-level standardized template in NX modeling software, or the modeling process can be replaced with a program developed in C# for the second time to form a general and expandable structure parameterized model library. On the basis of the central structured data model library and parts-level standardized model library, an expandable.exe script file program for automated modeling of the aero engine scheme design stage structure on the basis of central structured data is developed. As shown in Fig. 13, the entire process of automated implementation of a turbofan engine's overall structural 3D modeling driven by the central structured data XML semantic model file takes only 10 m.

Structural Information Extraction and Reconstruction Based on the Central XML Semantic Model

Currently, in the detailed design stage, geometric design models are usually CAD models drawn using NX software, with precise mathematical parameters essentially used to calculate geometric information. Rendering aero engine overall level data requires considerable time, and meeting real-time interaction requirements during the design review process is difficult. The specific requirement is to transform real structural information into lightweight models, develop a general interface that adapts to different engine structures to automatically complete the above work, and extract effective structural information, including the entire machine BOM structure, spatial position, placement posture, links, material properties, volume, and mass, into the central XML semantic model.

Through the development of automated interface scripts, an interaction between the central structure language XML file and the NX overall structure model is established. Using the object-oriented C# programming language for secondary development of NX modeling software, the engine BOM structural information is extracted and exported to an XML file, and the same parts models are merged to form a basic model library, as shown in Fig. 14. To ensure the consistency and traceability of the structure and multidisciplinary simulation data, a central structure XML semantic expression model (Fig. 15) is constructed. The BOM structure, spatial positions, placement postures, links, expressions, material properties, volumes, masses,

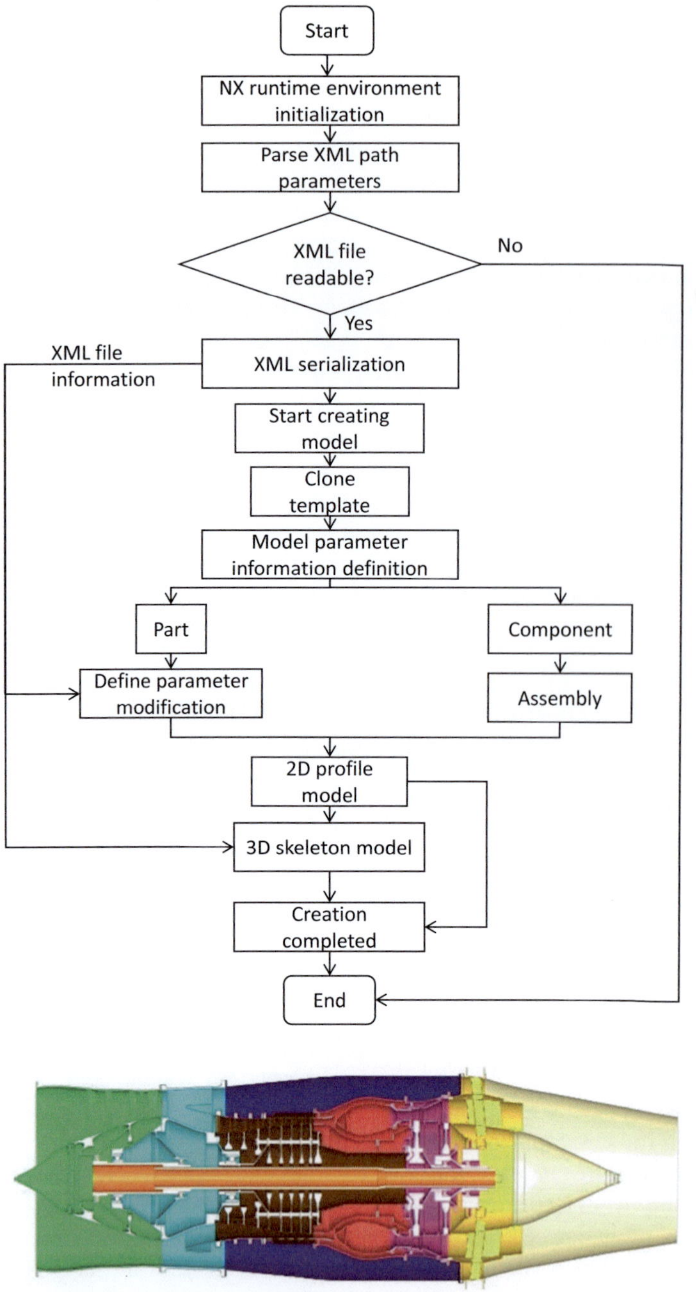

Fig. 13 XML model-driven automated modeling program execution process

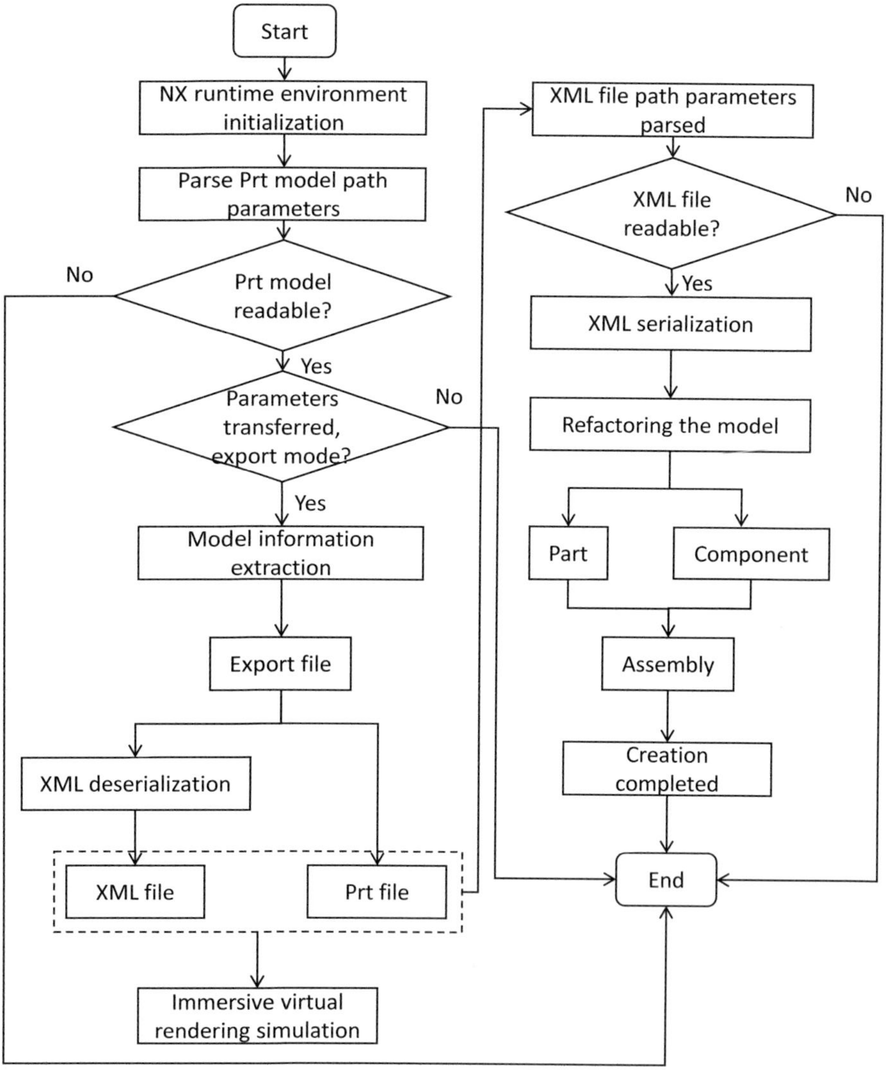

Fig. 14 Automated lightweight mapping process driven by the entire machine assembly model

moments of inertia, etc., of the engine's overall model parts are extracted and exported to an XML file to provide a data source for subsequent lightweight model automated reconstruction based on the XML semantic model.

The general engine BOM structure export and reconstruction script can realize the reconstruction of and information extraction from any engine structure model. It can automatically capture the different installation positions of the engine assembly parts in multiple positions and merge them, implement local coordinate and global coordinate conversion and mapping relationship technology, realize the lightweighting

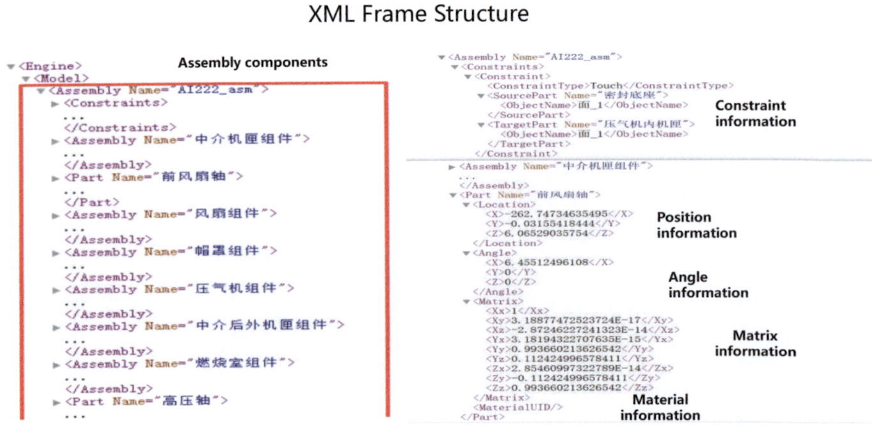

Fig. 15 Semantic model of structural information that is based on the entire machine assembly model

of structural data, and achieve a realistic, smooth immersive expression effect of real-time mapping of structural data. Currently, the automated iteration time for a certain turbofan engine is within 1 h.

Conclusions

The digital transformation of aero engines requires breaking down the information barriers throughout the entire lifecycle of R&D processes, building authoritative data sources and integrated platforms for engine products, and forming through-line automated, digitally driven thread. This article provides a detailed analysis of the research backgrounds and current statuses of digital thread-related concepts, methods, sources, and technical implementations in China and other countries. It proposes a technical route for using a feature component library assembled on the basis of metadata standards to create authoritative data sources on the basis of a central semantic model. It also suggests establishing semantic mapping models for various stages and disciplinary professional models and using automated mapping universal interfaces for different professional models to achieve interaction, and it further proposes an aero engine digital thread platform architecture. Taking the specific application scenario of CAD structural data penetration as an example, research and practice have been carried out from two aspects, namely, the structural expression and modeling of the central XML semantic model and the extraction and reconstruction of structural information on the basis of the central XML semantic model, thus providing a theoretical reference for the construction and application of a digital thread in industry.

References

1. Ferrari A, Nagel A (2020) Connecting, tracing and managing the lifecycle of models, simulation and linked data: is that easy?. Raytheon Technologies
2. Vosgien T, Van TN, Jankovic M et al (2012) Towards model-based system engineering for simulation-based design in product data management systems. IFIP Adv Inf Commun Technol 388(2):612–622
3. Singh V, Willcox K (2018) Engineering design with digital thread. AIAA J 56(11):4515–4528
4. Blackburn DM et al (2021) Transforming systems engineering through model- centric engineering. NAVAIR SERC
5. Dunbar D, Hagedorn T, Blackburn M et al (2023) Driving digital engineering integration and interoperability through semantic integration of models with ontologies. Syst Eng 26(4):365–378
6. Zhang M, Jungo A, Gastaldi AA et al (2018) Aircraft geometry and meshing with common language schema CPACS for variable-fidelity MDO applications. Aerospace 5(2):47
7. Hedberg T, Lubell J, Fischer L et al (2016) Testing the digital thread in support of model-based manufacturing and inspection. ASME J Comput Inf Sci Eng
8. Medani O, Ratchev SM (2006) A STEP AP224 agent-based early manufacturability assessment environment using XML. Int J Adv Manuf Technol 27(9–10):854–864
9. Khurana P, Sormaz DN, Khetan R (2006) Integration of CAD, CAPP and process modeling using XML technologies. In: MSEC2006. Manufacturing science and engineering, parts A and B, pp 539–546
10. Wang JY, Mo R (2012) Studies on process information expression techniques of 3D CAPP. Appl Mech Mater 121:2258–2262
11. Dai G, Zhang Y, Wang H et al (2004) Research on the system architecture of digital pre-assembly of products. Comb Mach Tool Autom Mach Technol (7):39–41
12. Jianguo C (2018) Research status, challenges and prospects of aero engine simulation technology. Propuls Technol 39(5):961–970
13. Kwon S, Monnier LV, Barbau R et al (2020) Enriching standards-based digital thread by fusing as-designed and as-inspected data using knowledge graphs. Adv Eng Inform 46:101102
14. Zhu Y, An Y, Zi Y et al (2022) IoTDataBench: Extending TPCx-IoT for compression and scalability. In: Nambiar R, Poess M Performance evaluation and benchmarking. Cham: Springer International Publishing, pp 17–32

Cross-dimensional Model on Overall Performance Calculation Method of the Turbofan

Abstract This chapter proposes a multi-dimensional turbofan engine overall performance calculation method that is coupled with a 0D/3D simulation model, which considers the effect of component geometry on aerodynamic performance to improve the accuracy of overall engine performance prediction. Using a mixed exhaust turbofan engine as the application object, on the basis of a zero-dimensional overall performance simulation program framework, the fan component and nozzle component in the overall simulation program are replaced by a tightly coupled method to obtain the 0D/3D multi-dimensional overall engine performance simulation results. The physical mechanism of the effect of the three-dimensional model on the overall performance is analyzed. The performance simulation accuracy under different-dimensional model mixing situations is compared with the experimental test data. Compared with the zero-dimensional model, 0D/3D coupling significantly improves the calculation accuracy for physical quantities at key locations, such as the turbine inlet temperature, and the simulation accuracy for performance parameters such as the overall engine thrust and fuel consumption rate can be improved by 2%. The use of component performance models of different dimensions for multi-dimensional simulation can effectively improve the overall aerodynamic performance simulation results for the whole engine, and the balance between calculation efficiency and simulation accuracy can be flexibly adjusted according to needs.

Keywords Turbofan engine · Overall performance · Cross-dimensional · Tight coupling · Three-dimensional simulation model

Introduction

To realize the three major features of digital engine system simulation, interconnection, and dynamic evolution, a high-precision engine overall performance model must be established. The establishment of a high-precision overall performance model requires accurately predicting the characteristic curves of each component of the engine, accurately obtaining the matching relationship between each pair of

Q. Xu et al., *Digital Intelligent Aeroengine*,
https://doi.org/10.1007/978-981-95-0942-3_5

components, accurately recording the model updates during the operating stage, and providing a basis for the prediction of overall performance by engine R&D and user departments. Currently, high-precision overall performance models are developing toward higher dimensions and precision.

Zero-dimensional simulation relies on component characteristic databases, with data from experiments and experience. When new engine models are developed, accurate characteristics of the components are lacking, which makes accurate prediction of the overall performance parameters using zero-dimensional simulation models difficult. By introducing a three-dimensional simulation model, more accurate component characteristics can be obtained on the basis of the geometries of new components, thereby improving the prediction accuracy for the overall performance parameters. In addition, compared with low-dimensional models, three-dimensional simulation models introduce fewer assumptions, simplifications, and empirical rules; have higher accuracy in predicting component performance; and require more calculations. With the development of high-performance computing technology, applying three-dimensional simulation models of turbofan engine components to overall performance calculations is now possible. When a zero-dimensional/three-dimensional multi-dimensional model is used, both the calculation efficiency and accuracy can be considered.

This paper proposes an overall performance calculation method that is tightly coupled with the component "zero-dimensional/three-dimensional multi-dimensional model". This study analyzes the common working constraints that need to be satisfied for multi-dimensional model overall performance simulation, constructs the corresponding common working equation group, and studies the physical field information transmission technology between zero-dimensional component models and three-dimensional component models. By directly replacing the zero-dimensional models of the fan and nozzle components with three-dimensional models, based on the improved Newton–Raphson method, iterative calculations are performed to obtain the flow rate, speed, power, efficiency, pressure ratio, and other component parameters of the turbofan engine on the common working line, as well as the thrust and fuel consumption rate performance parameters of the whole engine.

Preliminary Discussion

The numerical propulsion system simulation (NPSS) program, which is conducted in the United States, is a famous project in the field of propulsion system numerical simulation [1–6] that aims to accelerate the implementation of the integrated high-performance turbine engine technology (IHPTET) program by efficiently utilizing advanced digital information and communication technology [7–9]. In the versatile affordable advanced turbine engine (VAATE) program [10], a numerical simulation system was developed and applied as depicted in Fig. 1. Under the promotion of the NPSS program, American researchers have continuously developed and improved the numerical simulation system of aero engines, successfully created a multidisciplinary

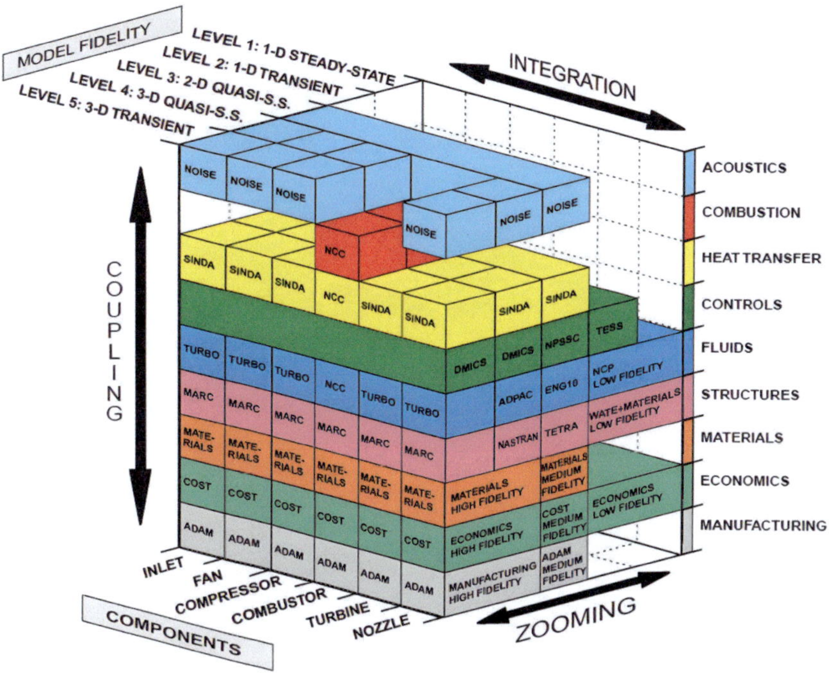

Fig. 1 NPSS overall engine multi-dimensional calculation

simulation platform, and conducted in-depth numerical simulation analysis of many engines under research and in service. The NPSS platform is a typical representative platform with increasing dimensions.

Turner [11, 12] and Lytle [13] conducted numerical simulations of the GE90 engine's overall flow field. As shown in Fig. 2, Turner's full engine simulation results illustrate the overall flow structure, but when analyzing the overall simulation results, they reported that there were still deviations from the cycle state point parameters. In addition, researchers from Stanford University used the coupler for high-performance integrated multiphysics simulations (CHIMPS) software platform to simulate a Pratt & Whitney (PW) engine. A numerical simulation was conducted on the whole engine flow field of a turbofan engine from Pratt and Whitney [14]. They combined RANS and LES solvers to simulate the flow field under cruising conditions and compared the numerical simulation results with experimental measurement results. The comparison results show that the spatial distributions of the interface parameters of the three subsystems have different degrees of impact on the performance of the components on both sides of the interface (Fig. 2).

Compared with other countries, although China has had a lower starting point and a later start in this field, it has still conducted a large amount of research. Most of the research has focused on characteristic simulations, and numerical simulation technology for components has reached quasi-three-dimensional and even

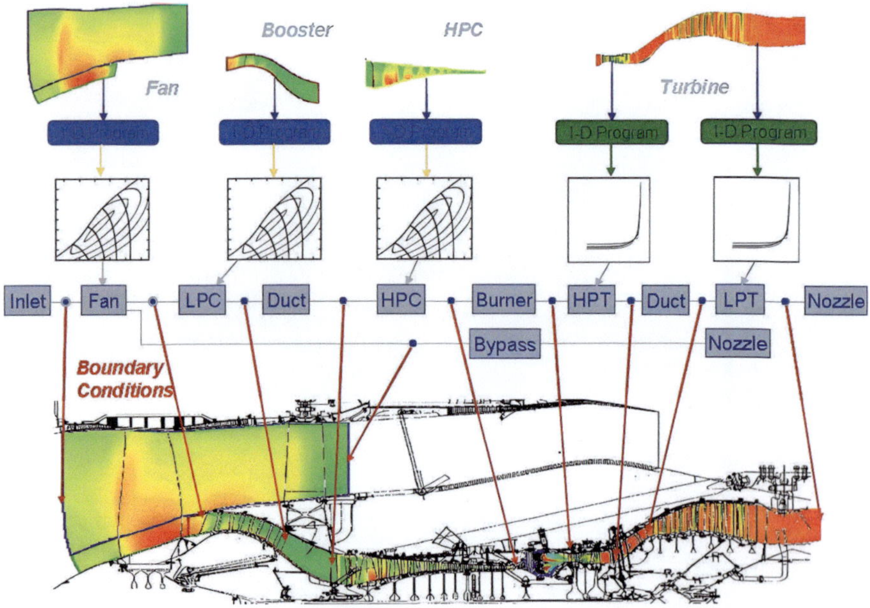

Fig. 2 Turner's full engine simulation results for the GE90 engine

three-dimensional accuracy. However, in terms of multidimensional comprehensive numerical simulations of whole engines, China's research level is still significantly behind the international advanced level. At the beginning of this century, China's scientific research institutions and universities began to develop a series of engine simulation programs [15]. These programs have borrowed the program structures of DYNGEN [16] and GENENG [17, 18], realized the coupled calculation of component characteristics, and made adjustments on the basis of experimental results, thus optimizing the empirical model.

Since the 1990s, China has made many advances in the field of aero engine simulation, but these advances have focused mainly on real-time control [19], component characteristic correction and other related fields [20, 21]. In terms of aerodynamic calculations, progress has been made mainly around the three core components of the aero engine: the compressor, turbine and combustion chamber. With the growing demand for overall simulation and modeling of propulsion systems [22], quickly and accurately implementing simulation technology for the whole aero engine has become a key problem to be solved.

For the dimensionality of the aero engine simulation model, this article adopts the following definition of Ivanov [23]:

(1) A zero-dimensional model uses charts, approximate formulas, statistical data, interpolation algorithms, etc., to calculate engine characteristics and is used to describe the relationships between subsystem components and overall performance. A one-dimensional model, on the average radius of the engine flow

path, establishes mass, momentum and energy conservation in one geometric dimension. It includes one-dimensional steady-state and one-dimensional transient models. The cascade method is often used; the advantage of this method is its fast calculation speed, and the disadvantage is that the simplification in this method easily leads to a lack of accuracy.

(2) A two-dimensional model, which is also known as a quasi-three-dimensional model, is based on the two types of flow surface theory that were proposed by Professor Wu Zhonghua in 1952. In the engine flow path, the three-dimensional flow field is decomposed into two perpendicular two-dimensional flow surfaces to decouple and interactively solve via a solution method that involves the radial balance equation, the Euler equation or the N–S equation.

(3) In a three-dimensional model, according to the engine components and three-dimensional flow path structure, a full three-dimensional internal flow path can be selected for detailed aerodynamic and thermal calculations. Because the structure is complex, the number of nodes in the calculation grid for a three-dimensional model usually exceeds 10e6, and the number of iterations exceeds 10e3, which depends more on the floating-point computing performance of the computer. Moreover, the calculation consumes much of the CPU power and memory of the computer.

In summary, with the exception of the zero-dimensional model, the models essentially have different dimensions and use fluid equations to describe the reduced dimensions of the flow field. The zero-dimensional model uses more experimental external characteristics of the components to establish the engine model, does not consider the internal flow fields of the components, requires less calculation and has a fast operating speed; the main disadvantages of the zero-dimensional model are that it lacks a description of the internal parts of the components and needs experimental data correction. For the preliminary design or engine maintenance stage, if the component characteristic map is accurate, the calculation of the zero-dimensional program is sufficiently accurate. However, in actual simulation calculations, the characteristic maps of all the components are not always available, such as when a component is modified, in the design stage where there is no characteristic map, when the experimental testing cost is too high, or when the characteristic map cannot be obtained for various reasons. A zero-dimensional program that depends completely on the characteristic map has difficulty meeting the research needs in terms of accuracy. When there are minor changes in the components, retesting the new characteristic map requires considerable time and cost.

Calculation Principles and Methods

Zero-Dimensional Modeling Method

A whole-engine overall performance simulation calculation program is developed on the basis of zero-dimensional models, and a calculation flow chart is shown in Fig. 3. On the basis of the aerodynamic and thermal principles of turbofan engine components, zero-dimensional aerodynamic models are established for the fan, high-pressure compressor, combustion chamber, high-pressure turbine, low-pressure turbine, mixing chamber, nozzle, etc. On the basis of the constraint relationships of internal and external flow balance and high- and low-pressure shaft power balance, a common working equation group is constructed with the input parameters of the component model. The program iteratively solves the common working nonlinear equation group by obtaining reasonable initial value ranges of input parameters such as the low-speed shaft speed, the high-speed shaft speed, the fuel flow, the nozzle throat area, and the variable guide vane angle via the classic Newton–Raphson method and obtaining the flow, thrust and fuel consumption rate of the whole engine under the given working conditions; the aerodynamic parameters of the key section; and the performance results, such as the pressure ratio, efficiency, and power of the rotating components. The complete zero-dimensional model engine performance calculation process is shown in the following figure, where the engine is composed of the following component models: a fan characteristic map (FAN map), a high-pressure compressor characteristic map (HPC map), a zero-dimensional model of the combustion chamber (BURN model), a high-pressure turbine characteristic map (HPT map), a low-pressure turbine characteristic map (LPT map), a zero-dimensional model of the mixing chamber (MIX model), a zero-dimensional model of the nozzle (NOZ model), and a zero-dimensional model of the outer duct (DUCT model).

In the zero-dimensional model program, there are 9 common working equations for the turbofan engine: the fan flow equation, compressor flow equation, high-pressure turbine equation, low-pressure turbine equation, mixing chamber inlet static pressure equation, mixing chamber back pressure equation, nozzle flow equation, low-speed shaft power equation, and high-speed shaft power equation. The common working equations that satisfy the flow balance and power balance of the above components are arranged in the following form:

$$\begin{cases} e_1 = f_1(x_1, x_2, \ldots, x_n) \\ e_2 = f_2(x_1, x_2, \ldots, x_n) \\ \vdots \\ e_n = f_n(x_1, x_2, \ldots, x_n) \end{cases}, n = 9 \tag{1}$$

$$error = F(x) = (f_1(x_1), f_2(x_2), \ldots, f_n(x_n))^T = 0 \tag{2}$$

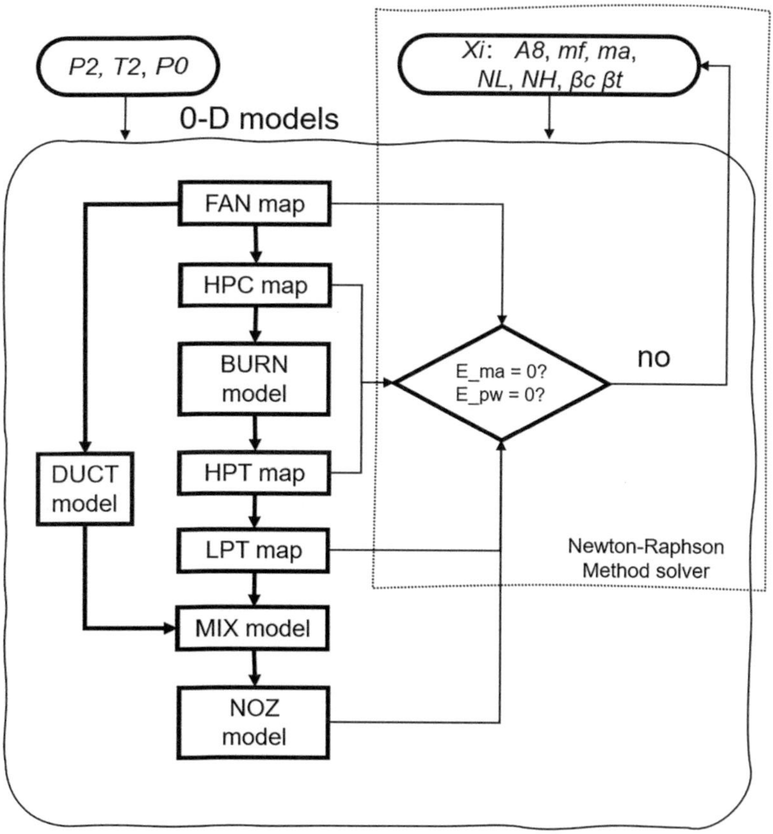

Fig. 3 Zero-dimensional model performance simulation calculation process

where the vector to be solved x is the input parameter of the zero-dimensional model when all parts meet the common working constraints, and the number of independent variables n in this method is 9.

$$x = (x_1, x_2, \ldots, x_n)^T \tag{3}$$

The common working equation can be rewritten as

$$\sum_{j=1}^{n} \frac{\partial f_i\left(X^{(k)}\right)}{\partial x_j} \delta_j^{(k)} = f_i\left(X^{(k)}\right), \quad i = 1, 2, \ldots, n \tag{4}$$

The left side of the above equation can be rewritten as

$$F'\left(x^k\right) = \left(\frac{\partial f_i(x)}{\partial x_j}\right)_{n\times n} \tag{5}$$

$$\frac{\partial f_i\left(X^{(k)}\right)}{\partial x_j} = \frac{f_i\left(X_j^{(k)}\right) - f_i\left(X^{(k)}\right)}{h} \tag{6}$$

The following Newton–Raphson iteration formula is constructed for solving the nonlinear common working equation group:

$$x^{(k+1)} = x^{(k)} - \left[F'\left(x^{(k)}\right)\right]^{-1} F\left(x^{(k)}\right), \quad k = 0, 1, 2, n \tag{7}$$

For the zero-dimensional simulation model of engine performance, the component-level modeling method is generally used; that is, the thermodynamic model of each component is calculated in the order of the gas path, but the special feature of the engine model is that the components before and after the gas path are coupled together, and the input parameters of a single component model usually depend on the results output by other component models. To solve this noncausal modeling problem, a set of guess variables needs to be given in advance as the inputs of each component model, and then a set of residuals is calculated. The residuals are solved iteratively so that they are equal to zero. The equation that corresponds to the residuals is also called the common working equation.

Three-Dimensional Modeling Method

The aerodynamic performance of all the components of the turbofan can be obtained through three-dimensional CFD simulation calculations. Since the components are connected directly through the calculation grid, the flow field simulation calculation automatically satisfies the flow balance after convergence, so the performance simulation calculation of the complete three-dimensional model needs only to satisfy the power balance of the high- and low-pressure shafts by iteratively solving for control parameters such as fuel flow and speed. The complete three-dimensional model engine performance calculation process is shown in Fig. 4, where the engine is composed of three-dimensional simulation calculation models of the following components: the fan (FAN CFD), high-pressure compressor (HPC CFD), combustion chamber (BURN CFD), high-pressure turbine (HPT CFD), low-pressure turbine (LPT CFD), mixing chamber (MIX CFD), nozzle (NOZ CFD), and outer duct (DUCT CFD).

The complete three-dimensional model engine performance calculation has the fewest parameters to be solved for and fewest common working constraint equations but requires the longest time for component performance calculation and has the highest demand for computing resources. Therefore, in scenarios where fast whole engine performance calculation is needed, a feasible compromise strategy is to use

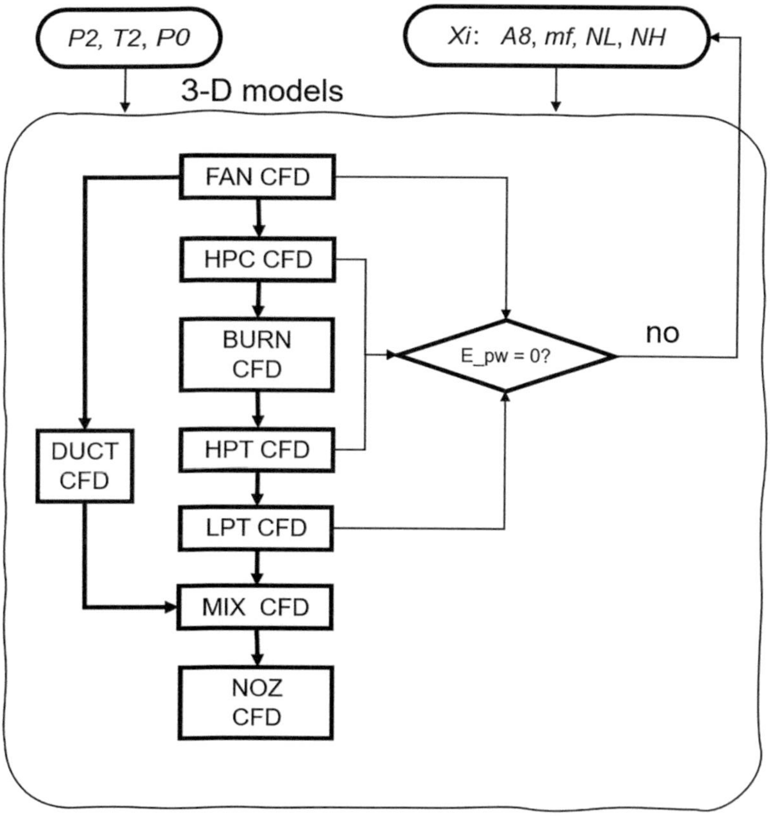

Fig. 4 Complete three-dimensional model performance simulation calculation process

three-dimensional models for various key components and zero-dimensional models for other components.

Zero-Dimensional/Three-Dimensional Modeling Method

The zero-dimensional/three-dimensional mixed multi-dimensional model calculation method is based on the zero-dimensional model calculation method and replaces various components with three-dimensional simulation models to improve the accuracy of the overall performance calculation. Taking the replacement of the fan and nozzle three-dimensional models as an example, the performance simulation calculation process of the zero-dimensional/three-dimensional mixed multi-dimensional model is shown in Fig. 5. The figure shows that the inlet boundary conditions of the three-dimensional fan model are determined by the total temperature and total pressure of the incoming flow, the outlet boundary conditions are calculated according to the flow balance condition under common working constraint conditions, and the speed is determined after the common working iterative calculation converges. The inlet conditions of the three-dimensional model of the nozzle are determined by the output parameters of the zero-dimensional model of the mixing chamber, and the outlet back pressure is adjusted on the basis of the common working flow balance condition. The input total temperature and total pressure of the high-pressure compressor zero-dimensional model are determined according to the simulation results of the fan three-dimensional model. The input total temperature and total pressure of the mixing chamber zero-dimensional model are jointly determined by the simulation results of the outer duct three-dimensional model and the output parameters of the low-pressure turbine zero-dimensional model.

The boundary conditions of the three-dimensional model are given on the basis of the output parameters of the zero-dimensional model; the input parameters of the zero-dimensional model also depend on the simulation calculation results for the three-dimensional model, so designing the physical parameter statistics and transmission protocol between the inlet and outlet boundaries of different-dimensional models and developing corresponding code tools are necessary. Since the simulation calculation of the three-dimensional model takes a long time and there is a possibility of nonconvergence, the solution strategy for the common working equation of the multi-dimensional model needs to be adjusted compared with that of the zero-dimensional model to improve the convergence of the overall performance simulation calculation. This chapter introduces the multi-dimensional model data exchange

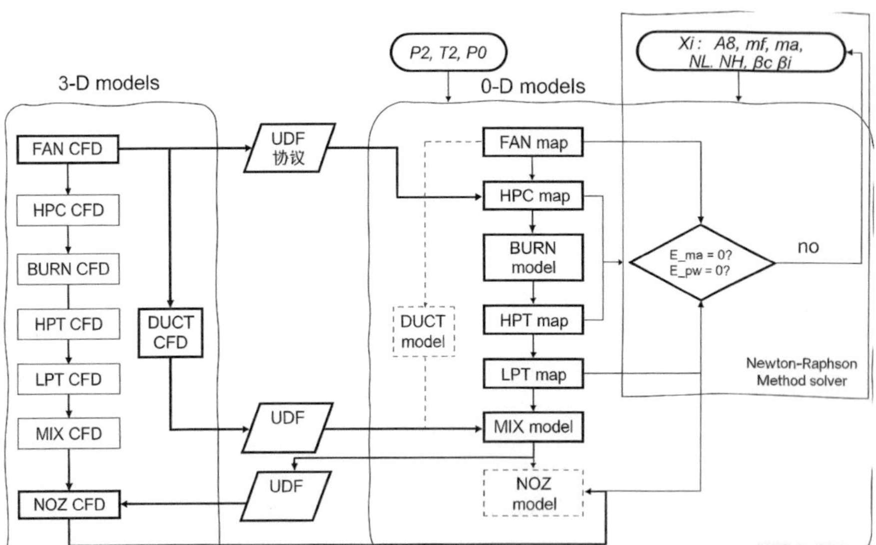

Fig. 5 Zero-dimensional/three-dimensional mixed multi-dimensional model performance simulation calculation process

strategy and multi-dimensional model iterative solution strategy used in the performance simulation calculation process of the zero-dimensional/three-dimensional mixed multi-dimensional model.

Multi-dimensional Model Data Transfer Strategy

In the multi-dimensional calculation process that is based on the zero-dimensional program framework, when flow field simulation software is used to calculate a certain component's three-dimensional model. The exit flow rate, total pressure, total temperature, etc., obtained from the zero-dimensional model calculation for the upstream component need to be input into the three-dimensional calculation software as the inlet boundary conditions of the three-dimensional model. After the three-dimensional model flow field simulation calculation is complete, the three-dimensional flow field results of the component outlet section are analyzed statistically, and the flow rate, total pressure, total temperature, etc., are obtained as the inlet input parameters of the downstream component through the dimension-reduction statistical method. The data transfer process at the interface between the zero-dimensional and three-dimensional models is shown in Fig. 6.

Therefore, extracting parameters such as the flow rate, total pressure, and total temperature at the inlet and outlet interfaces of the three-dimensional and zero-dimensional components, performing dimensionality increase or decrease

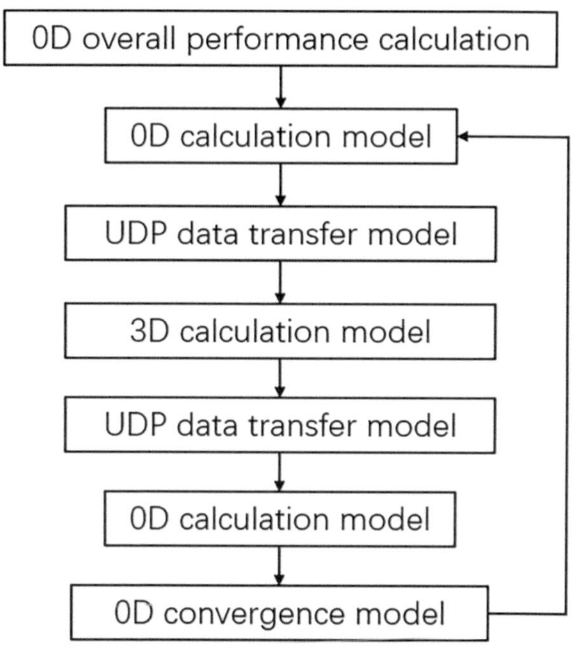

Fig. 6 Zero-dimensional/three-dimensional UDP data exchange process

processing, and passing the results to the upstream and downstream components are necessary. When the three-dimensional model passes data to the zero-dimensional model, an appropriate weighted averaging method can be selected to calculate the average physical parameters on the section. When the zero-dimensional model passes data to the three-dimensional model, the static pressure distribution along the radius can be obtained at the inlet of the three-dimensional model's flow channel according to the radial equilibrium equation, which is expressed as follows:

$$\frac{dp}{dr} = \rho \frac{c_t^2}{r} \tag{8}$$

where r represents the radius, p represents the static pressure at that radius, ρ represents the density at that radius, and c_t represents the circumferential speed at that radius. The radial distribution of the static pressure p can be calculated through the radial distribution of the circumferential speed c_t, which can be used as the boundary condition of the three-dimensional model.

In the iterative calculation process for the multi-dimensional model, data transfer between the zero-dimensional model program and the three-dimensional model simulation calculation program is realized by reading and writing specific UDP script files. The specific data transfer network communication process is shown in Fig. 7.

During the "overall performance calculation" process, the "network interface" receives the results output by the zero-dimensional model of the upstream engine component, sets the required boundary conditions of the three-dimensional model, passes the boundary conditions of the three-dimensional model to the three-dimensional calculation software through grid communication, and saves the results as a script file that uses the UDP protocol. After the "three-dimensional component calculation service" is started, the three-dimensional calculation software automatically reads this script file for the boundary assignment of the solver and starts the flow field simulation calculation. After the three-dimensional calculation of the

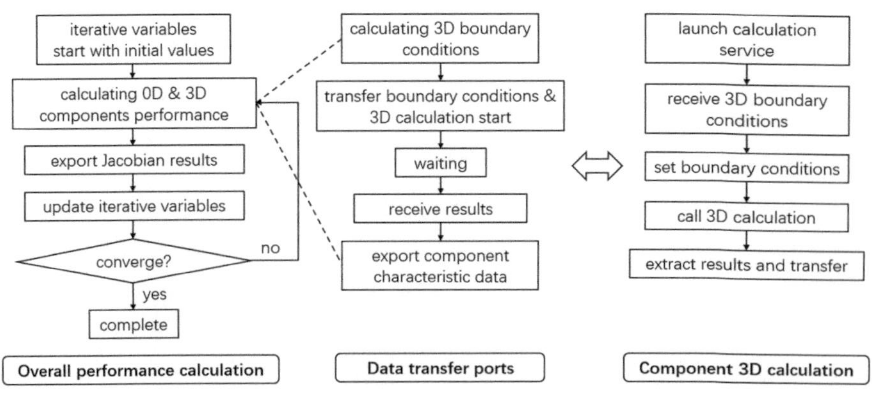

Fig. 7 Schematic diagram of data exchange network communication

component converges, the "three-dimensional component calculation service" statistically analyzes the physical parameters of the inlet and outlet boundary sections, passes them to the "network interface" through network communication, and then passes them back to the "overall performance calculation" main program through the "network interface" as the input parameters of other zero-dimensional component models.

Iterative Solution Strategy for a Multi-dimensional Model

In the process of multi-dimensional whole engine performance calculation based on the zero-dimensional program framework, the zero-dimensional models of components such as fans can be directly replaced by the flow field simulation of three-dimensional models. The total pressures and flow rates at the inlet and outlet of the fan components no longer depend on the interpolation of characteristic map data but are statistically extracted directly from the three-dimensional flow field results for the fan components. In the process of calculating the overall performance parameters of the whole engine, regardless of whether a zero-dimensional model or a three-dimensional model is used, iteratively solving the common working equation group by continuously adjusting the performance parameters of the components is necessary. Therefore, the solution process for the common working equation group based on the Newton–Raphson method is the same, and the 9 residuals in the common working equation group of the full zero-dimensional model are still used as the convergence judgment conditions.

The common working equation group of the mixed exhaust turbofan engine is established on the basis of the following balance constraint relationship:

(1) Fan (FAN) inlet and outlet flow balance

$$e_1 = \dot{m}_{FAN}(outlet) - \dot{m}_{FAN}(inlet) \tag{9}$$

(2) High-pressure compressor (HPC) inlet and outlet flow balance

$$e_2 = \dot{m}_{HPC}(outlet) - \dot{m}_{HPC}(inlet) \tag{10}$$

(3) High-pressure turbine (HPT) inlet and outlet flow balance

$$e_3 = \dot{m}_{HPT}(outlet) - \dot{m}_{HPT}(inlet) \tag{11}$$

(4) Low-pressure turbine (LPT) inlet and outlet flow balance

$$e_4 = \dot{m}_{LPT}(outlet) - \dot{m}_{LPT}(inlet) \tag{12}$$

(5) Mixer (MIX) inlet and outlet flow balance

$$e_5 = \dot{m}_{MIX}(outlet) - \dot{m}_{MIX}(inlet) \tag{13}$$

(6) Nozzle (NOZ) inlet and outlet flow balance

$$e_6 = \dot{m}_{NOZ}(outlet) - \dot{m}_{NOZ}(inlet) \tag{14}$$

(7) Mixer (MIX) outlet inner and outer duct static pressure balance

$$e_7 = p_{MIX}(bypass) - p_{MIX}(core) \tag{15}$$

(8) High-pressure shaft power balance: balance of power between the high-pressure turbine and high-pressure compressor

$$e_8 = P_{HPT} * \eta_{HPT} - P_{HPC} \tag{16}$$

(9) Low-pressure shaft power balance: balance of power between the low-pressure turbine and fan

$$e_9 = P_{LPT} * \eta_{LPT} - P_{FAN} \tag{17}$$

The residuals of the fan, high-pressure compressor, high-pressure turbine, low-pressure turbine, mixer, nozzle inlet and outlet flows, mixer outlet inner and outer duct static pressures, high-pressure turbine and high-pressure compressor powers, and low-pressure turbine and fan powers form the following residual nonlinear equation group:

$$\mathbf{e}(\mathbf{x}) = \begin{vmatrix} e_1 \\ e_2 \\ e_3 \\ e_4 \\ e_5 \\ e_6 \\ e_7 \\ e_8 \\ e_9 \end{vmatrix} \tag{18}$$

where $\mathbf{x} = (x_1, x_2, \cdots, x_9)^T$.

The independent variable $\mathbf{x}$ of the nonlinear equation group $\mathbf{e}(\mathbf{x})$ is composed of 9 components: the high-pressure shaft speed x_1, low-pressure shaft speed x_2, fan inlet flow x_3, bypass ratio x_4, fan pressure ratio x_5, high-pressure compressor pressure ratio x_6, high-pressure turbine pressure drop ratio x_7, low-pressure turbine pressure decrease ratio x_8, and mixer outlet static pressure x_9. If the residual $\mathbf{e}(\mathbf{x})$ of the iterative solution of the nonlinear equation group drops to a set target value, the multi-dimensional model overall engine performance calculation is completed.

The iterative solution process for the overall performance of the whole engine using the zero-dimensional/three-dimensional mixed multi-dimensional model is shown in Fig. 8. During the iteration process, the step size dx of the independent variable **x** is fixed to 0.0001. The initial guess for the independent variable **x** is given according to the design performance of the component, and the adjustable range of the independent variable is limited to 0~200% of the initial guess. If the residual **e**(**x**) cannot drop to the target value within a set number of iterations, the initial guess for the independent variable **x** is increased or decreased by 5%, and a new round of Newton–Raphson iteration calculations is carried out until the residual drops to the target value.

Multi-dimensional Calculation Case

The overall performance simulation method for the zero-dimensional/three-dimensional hybrid multi-dimensional model proposed in this article is adopted, and the AI222 dual-axis mixed-flow turbofan engine is used as the application object to carry out the full zero-dimensional model performance simulation calculation, the multi-dimensional performance calculation of replacing the three-dimensional model of the nozzle, and the multi-dimensional performance calculation of replacing the fan three-dimensional model. The overall configuration of the AI222 engine is illustrated in Fig. 9. The design point performance parameters of the AI222 engine are shown in Table 1. The design speed of the low-pressure shaft is 13,200 rpm, and the design speed of the high-pressure shaft is 18,330 rpm (Fig. 9).

The structures of the main components of the AI-222 turbofan engine are shown in Fig. 10. The fan adopts a 2-stage axial flow type, with a titanium alloy integral blade structure; the high-pressure compressor adopts an 8-stage axial flow type, with the first 3 stages of the stator blades being adjustable; the combustion chamber is an annular combustion chamber that uses film cooling and a low-pollution design, with 16 aerodynamic atomizing nozzles; the high-pressure turbine is a 1-stage axial flow-type turbine that uses directionally solidified hollow blades, convection, and film cooling; and the low-pressure turbine is a 1-stage high-load turbine, with guide vane cooling.

Multi-dimensional Simulation Example of a Nozzle Three-Dimensional Model

A zero-dimensional model of each component of the AI222 turbofan engine is constructed and embedded into the overall performance calculation program for the

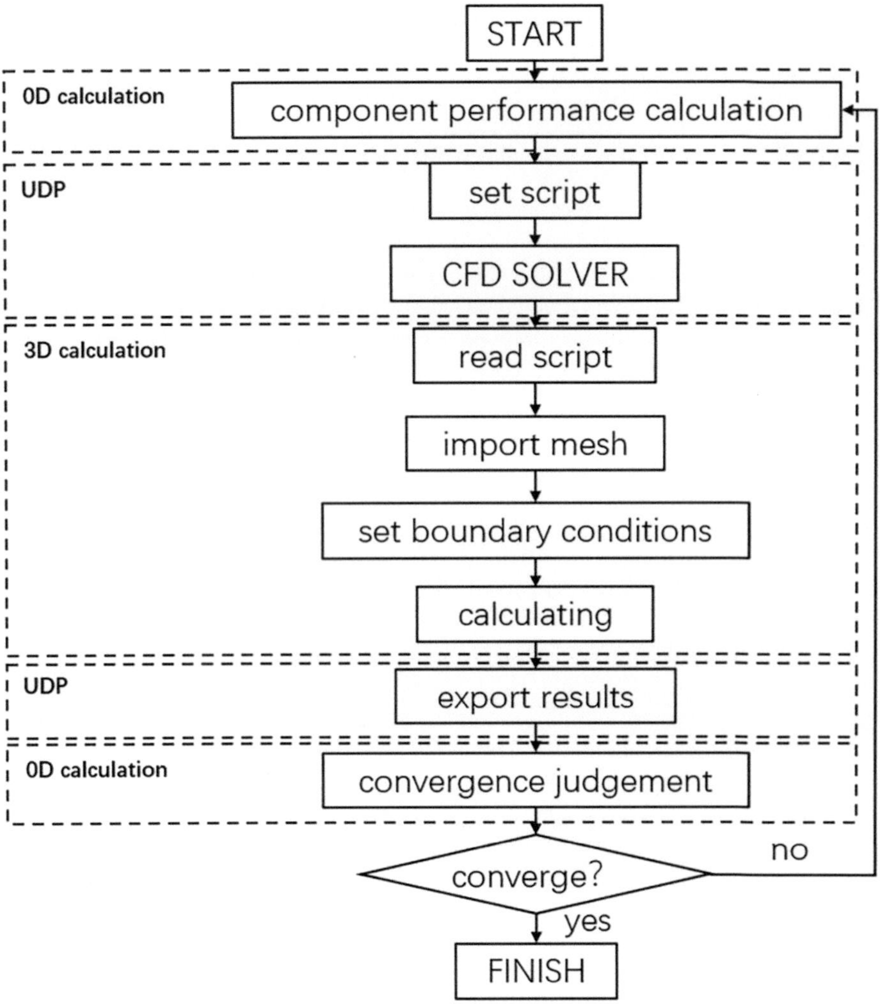

Fig. 8 Iterative solution process for the overall performance of the mixed-dimensional model

Table 1 AI222 design point overall performance test data

Engine model	АИ-222К-25
Compression ratio	15.4
Thrust-to-weight ratio	5.68
Air flow rate kg/s	49.4
Bypass ratio	1.18
Takeoff fuel consumption rate kg/(daN h)	0.651
Turbine inlet temperature K	1470
Takeoff thrust daN	2500

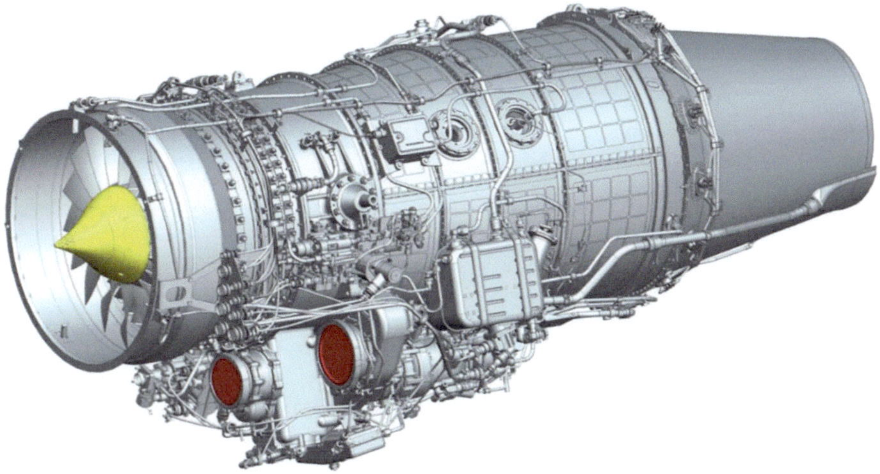

Fig. 9 AI-222 turbofan engine

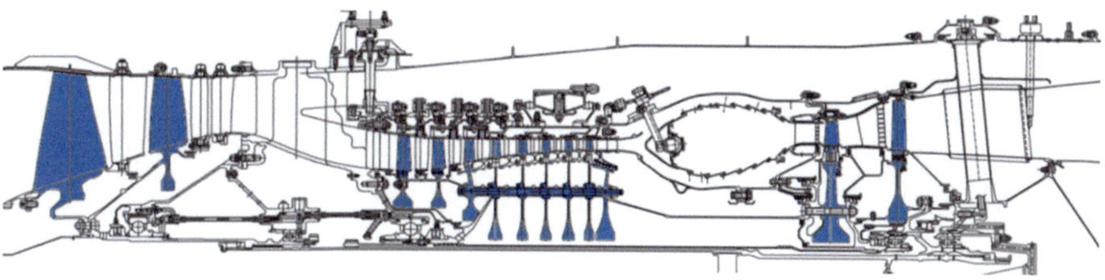

Fig. 10 AI-222 turbofan engine structure diagram

turbofan engine. A three-dimensional model of the nozzle component is constructed, as shown in Fig. 11, and the flow field mesh for the nozzle component is drawn, as shown in Fig. 12, while the calculation parameters are listed in Table 2. The three-dimensional calculation of the nozzle component uses a polygonal unstructured grid with 360,000 grid nodes; the total pressure and total temperature conditions at the outer duct inlet are given, the flow rate condition at the inner duct inlet is specified, and the static pressure condition at the nozzle outlet is established via the k–e turbulence model (Table 2).

The zero-dimensional model of the nozzle is replaced with a three-dimensional model, and the overall performance calculation for the whole engine is carried out. The convergence result is obtained after 11 h of calculation on an Intel Xeon Gold 6248R 48-core server. The Fig. 13 following figure shows a schematic diagram of the component calculation framework in this case. The flow rate and total pressure loss of the nozzle component are obtained directly by calculating the three-dimensional flow field with FLUENT software, and the performance of the remaining components is calculated using a zero-dimensional model.

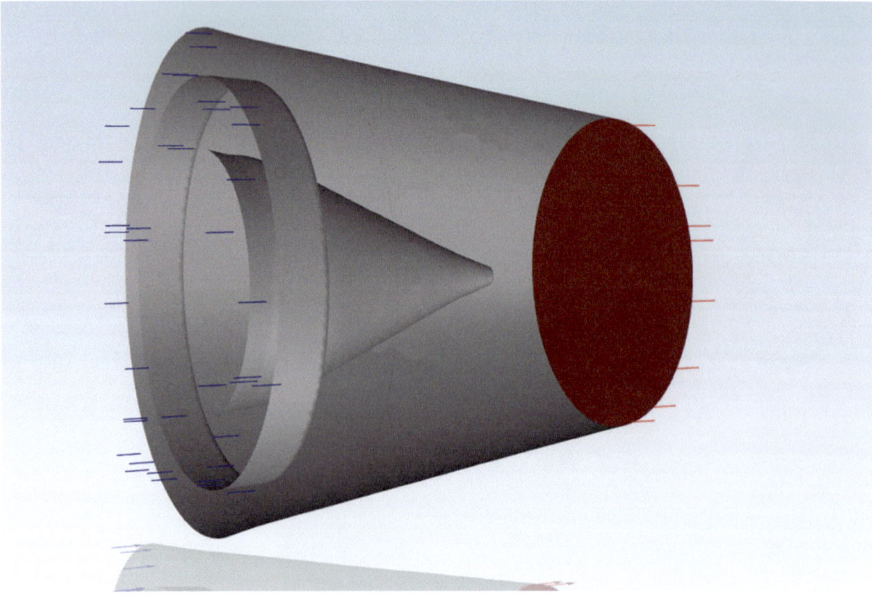

Fig. 11 Geometric model of the AI222 nozzle

Fig. 12 Cross-sectional view of the nozzle calculation grid

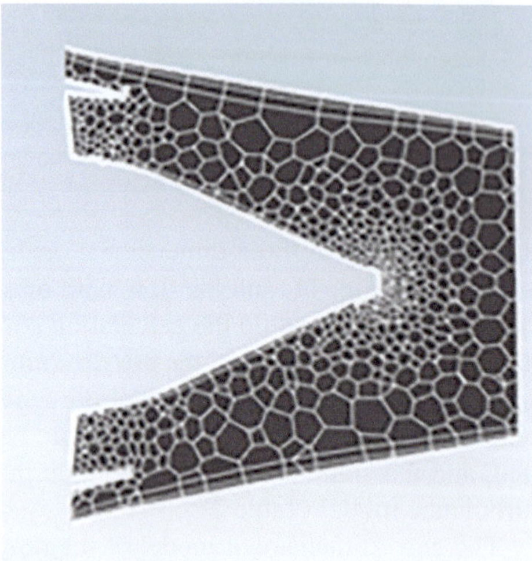

The distribution of the nozzle temperature on the meridian plane under the design point condition obtained by the computational fluid dynamics simulation software is shown in Fig. 14, and the pressure distribution on the meridian plane is shown in Fig. 15. The three-dimensional flow field distribution of the nozzle shows axial and radial nonuniformity, and the calculation accuracy of the total pressure loss and flow

Table 2 Nozzle component three-dimensional flow field calculation parameters

Grid type	Polygon
Grid node quantity	360,000
Turbulence model	Ke
Boundary conditions	Outer duct: total temperature and total pressure Inner duct: flow rate Nozzle outlet: static pressure
Calculation time	11 h
Computer hardware	Intel Xeon Gold 6248R 48 cores

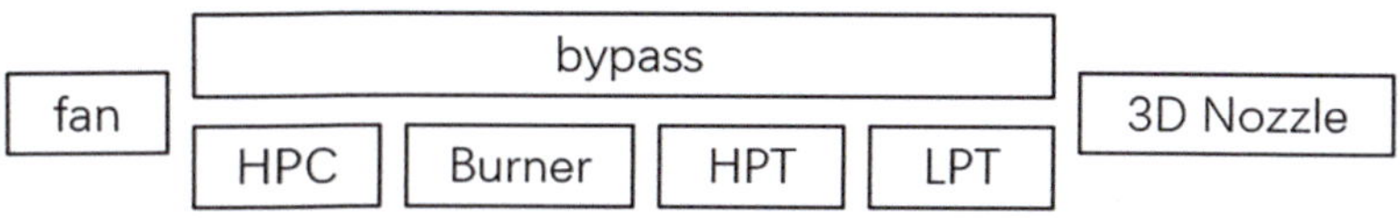

Fig. 13 Schematic diagram of the AI222 turbofan engine component calculation framework

rate is significantly improved compared with that of the zero-dimensional model. The trend of the axial velocity meridian plane distribution is not the same as that of the Mach number meridian plane distribution, especially near the outlet of the nozzle.

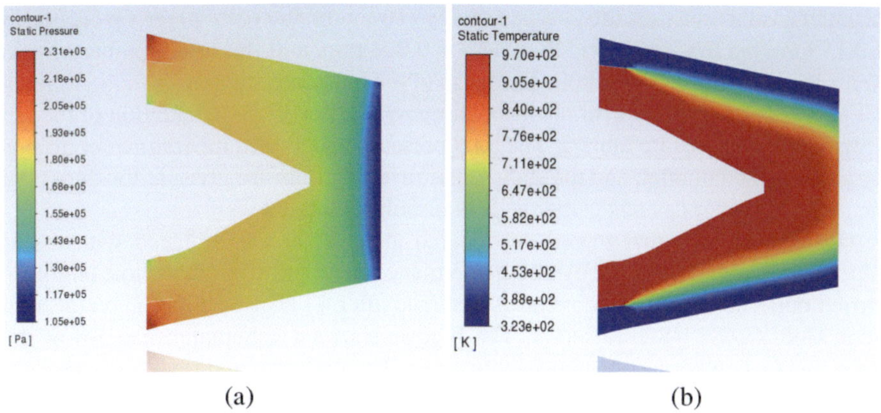

Fig. 14 Distributions of the static pressure and static temperature on the meridian plane of the nozzle at the design point

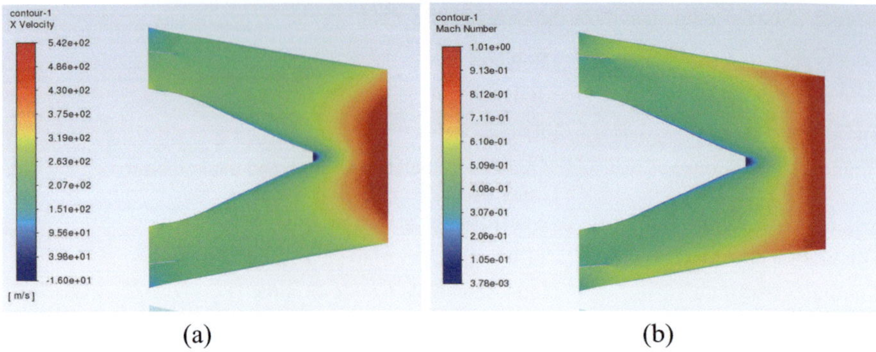

Fig. 15 Distributions of the axial velocity and Mach number on the meridian plane of the nozzle at the design point

Multi-dimensional Simulation Case of the Fan and Nozzle Three-Dimensional Model

A three-dimensional model of the fan duct component is constructed, as shown in Fig. 16. The fan grid is drawn in the graphical user interface of the AutoGrid5 module of NUMECA software. A single-channel computational domain hexahedral structured grid is used, and the rotor and stator flow path grids both use the O4H topology structure. The rotor blade tip clearance area uses the HO topology structure, the outer casing area uses the C-type topology structure, the rotor blade tip clearance is 0.15 mm, the first-layer grid thickness is 0.008 mm, and the total number of grid points is approximately 890,000. The mesh of the fan component flow field is shown in Fig. 17. The parameters for the three-dimensional flow field calculation of the fan component are listed in Table 3. The total pressure and total temperature conditions are given at the fan inlet, and the static pressure conditions are given at the outlets of the inner and outer casings using the S-A turbulence model.

The zero-dimensional models of the fan and nozzle are replaced with three-dimensional models, and the overall performance calculation for the whole engine is carried out. The convergence result is obtained after 44 h of calculation on an Intel Xeon Gold 6248R 48-core server. The Fig. 18 shows a schematic diagram of the component calculation framework in this case. The flow rate and total pressure ratio of the fan component are obtained directly by calculating the three-dimensional flow field with NUMECA software, and the flow rate and total pressure loss of the nozzle component are obtained directly by calculating the three-dimensional flow field with FLUENT software. The performance of the remaining components is calculated using a zero-dimensional model.

The design point operating conditions of the fan meridional plane static pressure and static temperature distribution obtained by the computational fluid dynamics simulation software are shown in Figs. 19 and 20. After boosting, the static pressure at the fan outlet cross-section is slightly greater than the inner casing static pressure,

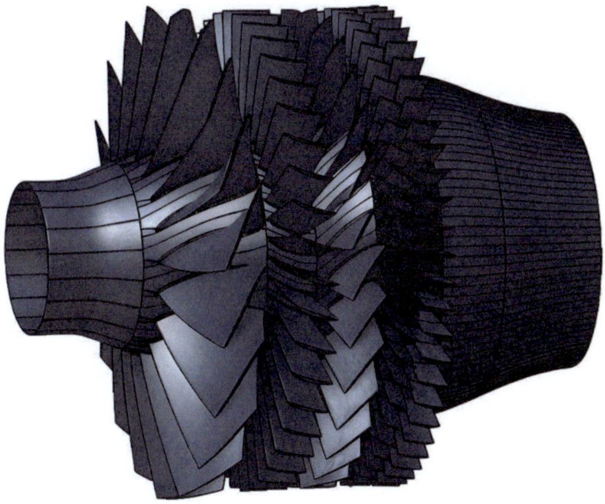

Fig. 16 Three-dimensional fan geometric model

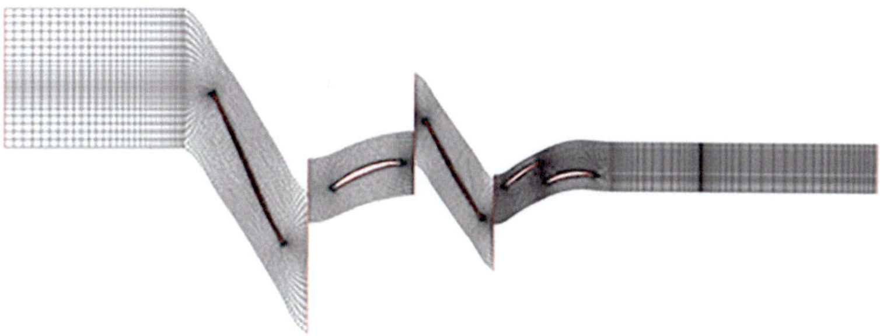

Fig. 17 Fan calculation grid

Table 3 Parameters for the three-dimensional flow field calculation of the fan component

Grid type	Structured hexahedron
Number of grid points	890,000
Turbulence model	SA
Boundary conditions	Fan inlet: total temperature and total pressure Outer casing outlet: static pressure Inner casing outlet: static pressure
Calculation time	44 h
Computer hardware	Intel Xeon Gold 6248R with 48 cores

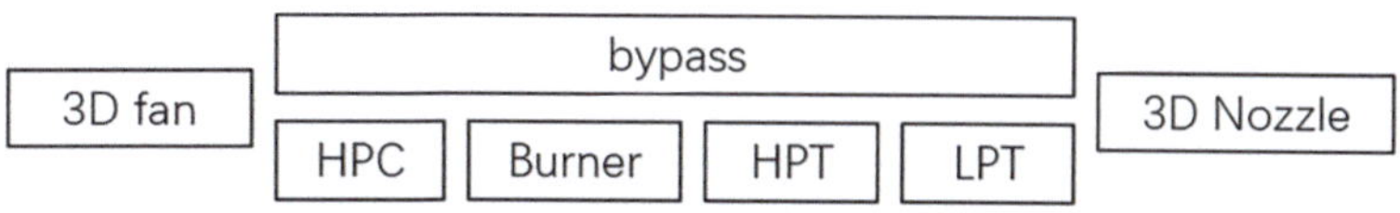

Fig. 18 Component calculation framework diagram

and the static temperatures of the inner and outer casings tend to be consistent. The calculation results of the static pressure and static temperature distribution at 50% of the blade height under the design point operating conditions, as shown in Figs. 21 and 22, reveal that there is a shock wave at the fan moving blade and that there is no obvious flow separation in the fan. The three-dimensional flow field simulation calculation yields more accurate fan inlet and outlet flows and total pressure ratios. When the fan blade geometry design changes, it can accurately reflect the impact of the geometric design on the component performance. After embedding the three-dimensional model of the fan component into the overall performance calculation process of the whole engine, it can further reflect the degree of influence of three-dimensional geometric optimization on the thrust and fuel consumption rate of the turbofan engine.

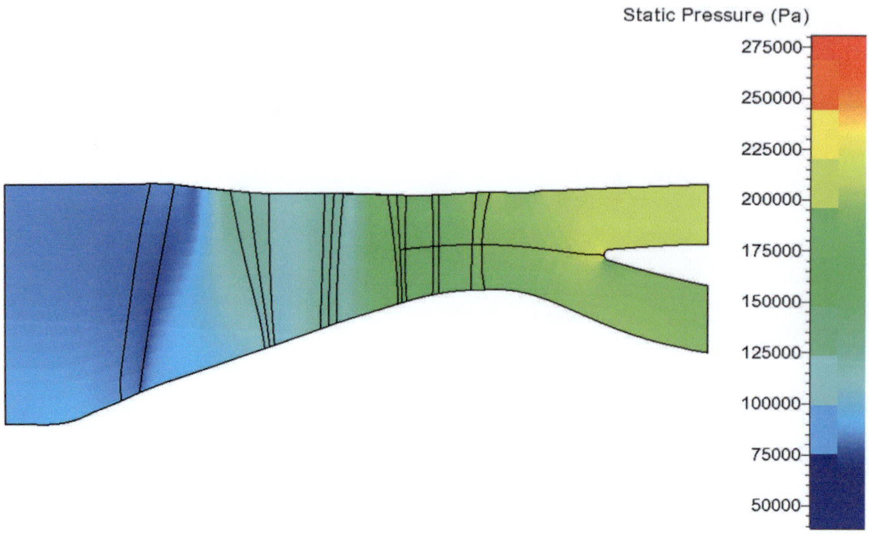

Fig. 19 Static pressure distribution of the fan meridional plane under design point conditions

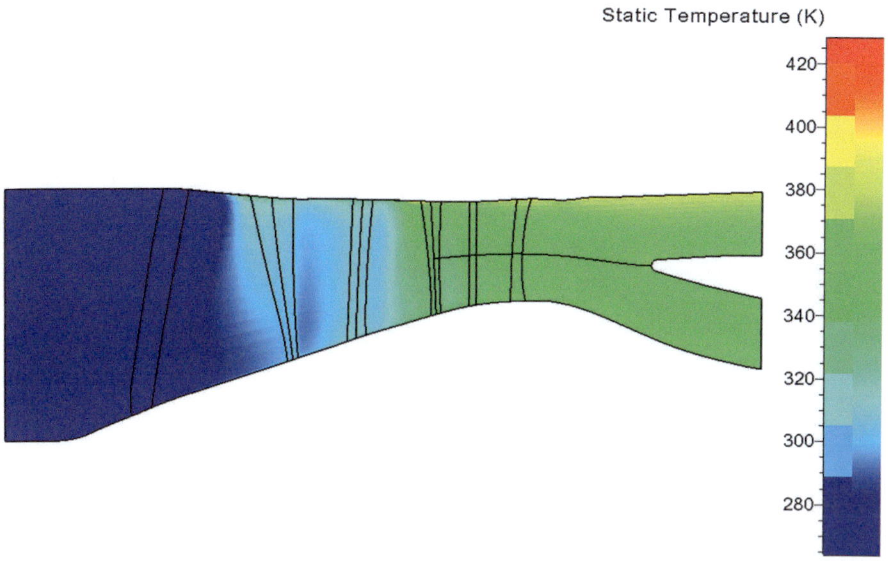

Fig. 20 Static temperature distribution of the fan meridional plane under design point conditions

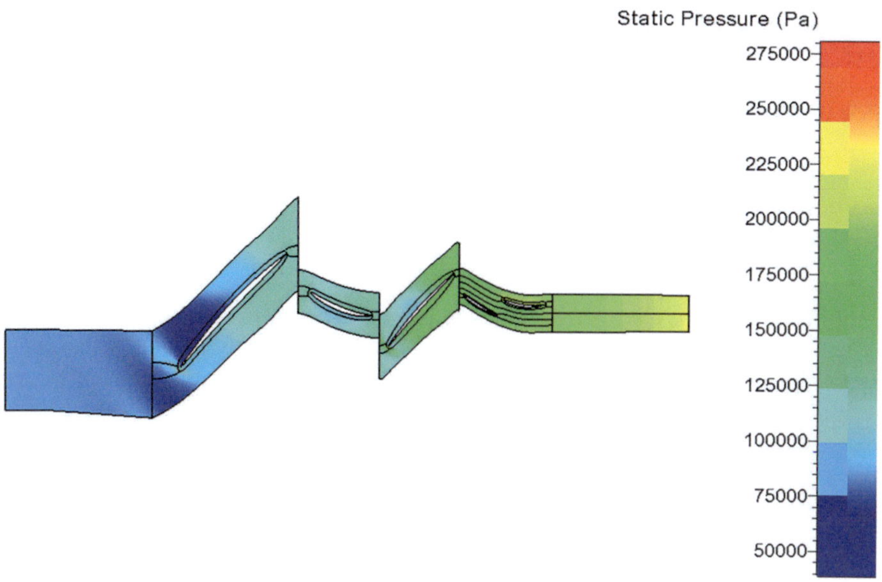

Fig. 21 Static pressure distribution at 50% of the blade height of the fan under the design point conditions

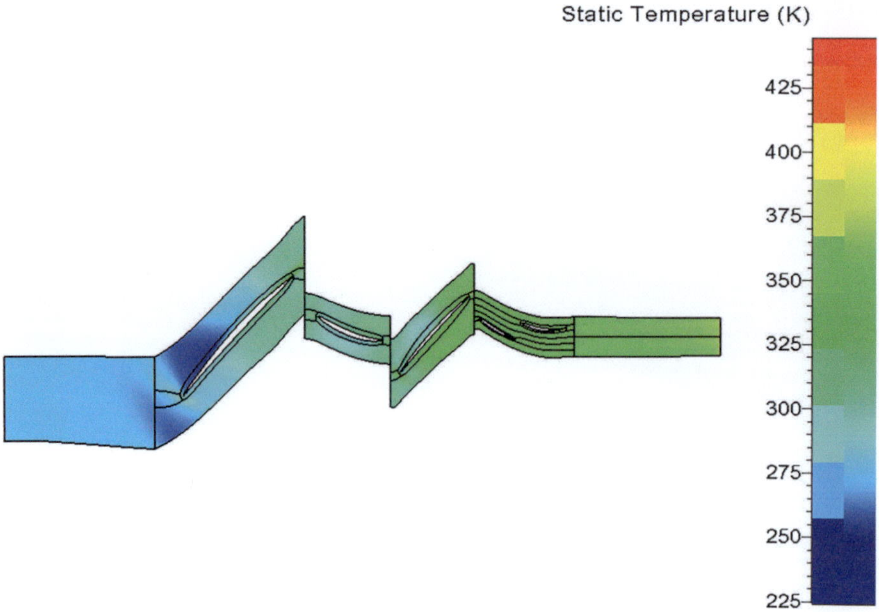

Fig. 22 Static temperature distribution at 50% of the blade height of the fan under design point conditions

Analysis of the Results

Comparison of the Calculation Accuracy for the Design Point Multi-dimension Model

Comparison of Fan Component Performance Parameters

By using the zero-dimensional/three-dimensional mixed multi-dimensional model overall performance simulation calculation method proposed in this paper, after replacing the fan component and nozzle component with a three-dimensional model, the performance parameters of each component of the AI-222 turbofan under the design point conditions were calculated and compared with the results obtained by the overall performance calculation of the full zero-dimensional model. Figures 23, 24, and 25 show comparisons of the fan component boost ratio, isentropic efficiency, and flow performance parameter values obtained by the multi-dimensional model and the full zero-dimensional model, respectively. Compared with the fan boost ratio obtained by the zero-dimensional model, the fan boost ratio obtained by the mixed-dimensional model calculation is lower, whereas the efficiency and flow rate are greater, thus reflecting the influence of the uneven axial, radial, and circumferential distributions of the fan component three-dimensional flow field, which results in

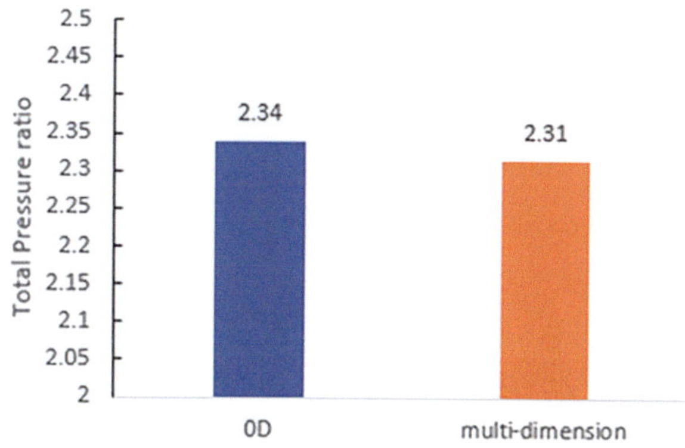

Fig. 23 Comparison of fan total pressure ratio between the zero-dimensional model and the multi-dimensional model

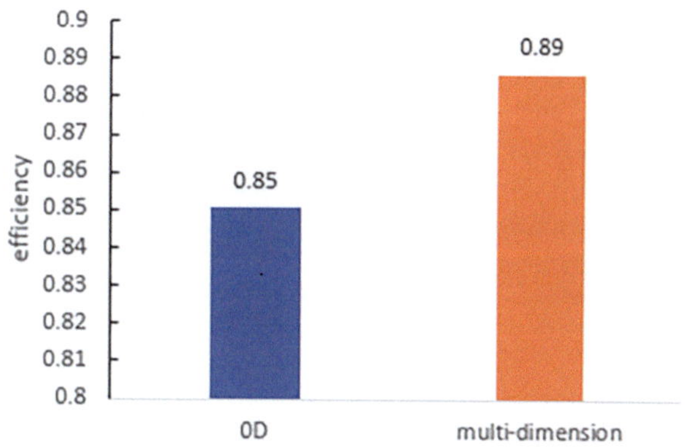

Fig. 24 Comparison of fan efficiency between the zero-dimensional model and the multi-dimensional model

the total pressure of the fan outlet cross-section three-dimensional flow field statistics being different from the total pressure obtained on the basis of the characteristic map interpolation. The fan three-dimensional model can reflect the effects of shock waves and nonuniformity on the flow field, and the calculation results for the flow rate, pressure ratio, and efficiency are more accurate.

Comparison of Errors in the Fuel Consumption Rate and Thrust Overall Performance Parameters

By using the zero-dimensional/three-dimensional mixed multi-dimensional model overall performance simulation calculation method proposed in this paper, after replacing the fan component and nozzle component with a three-dimensional model, the overall performance parameters of the AI-222 turbofan under the design point conditions were calculated, and the results were compared with the results obtained by the overall performance calculation of the full zero-dimensional model. Figure 26 and Table 4 show the errors of the whole engine air flow, bypass ratio, fuel consumption rate, and thrust parameters obtained by the zero-dimensional model calculation

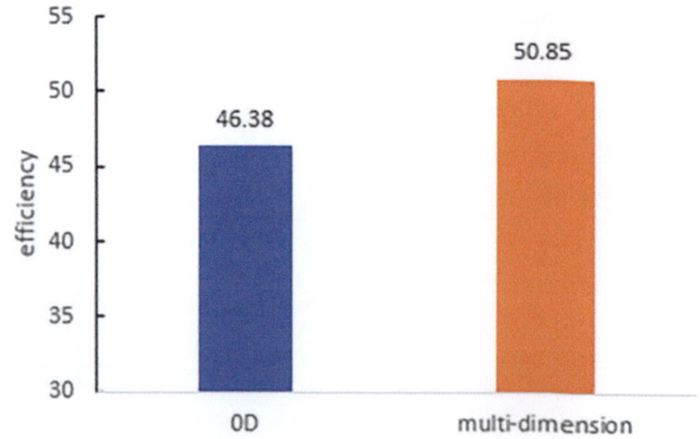

Fig. 25 Comparison of the fan flow rate between the zero-dimensional model and the multi-dimensional model

under the design point conditions and the design values, as well as the errors of the multi-dimensional model calculation results and the design values. The comparison results show that the calculation errors of the air flow, bypass ratio, fuel consumption rate, and thrust obtained by the multi-dimensional model are all smaller than those of the zero-dimensional model, thus fully reflecting the improvement in the calculation accuracy of the overall performance under the design point conditions by the multi-dimensional method.

Comparison of the Calculation Accuracy for the Multi-dimensional Model Performance at Off-Design Points

By using the zero-dimensional/three-dimensional mixed multi-dimensional model overall performance simulation calculation method proposed in this paper, after replacing the fan components and nozzle components with three-dimensional models, the overall performance parameters of the AI-222 turbofan engine working together under four kinds of fan low-pressure shaft speed conditions were calculated and compared with the results of the all-zero-dimensional model.

Figure 27 shows the trend of the zero-dimensional/multi-dimensional fan supercharging ratio working line. The fan supercharging ratio and flow rate increase with increasing speed, but compared with those of the zero-dimensional model, the supercharging ratio of the multi-dimensional model is generally lower, and the change is smoother with increasing speed, which is more in line with the actual working conditions of the fan components. Figure 28 compares the trends of the zero-dimensional and multi-dimensional fan efficiency working lines. The efficiency of the multi-dimensional model is generally higher than that of the zero-dimensional model and closer to the actual working efficiency of the fan. The comparison results for the fan pressure ratio and efficiency working line show that when the zero-dimensional model of the fan component is replaced with a three-dimensional model,

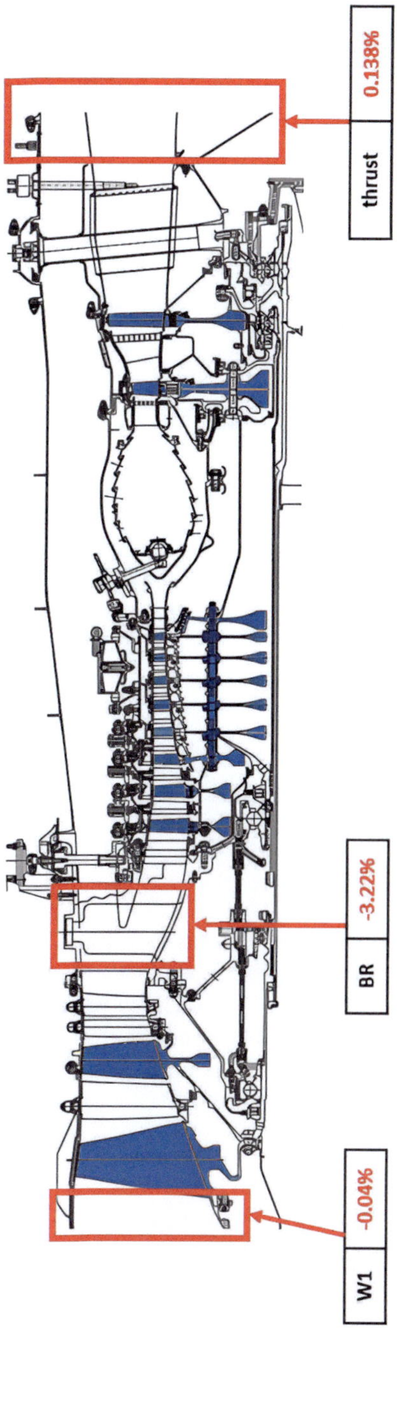

Fig. 26 Errors of different model results and design values of overall performance parameters under design point conditions

Table 4 Errors of the zero-dimensional/multi-dimensional model overall performance

Error %	Air flow	Bypass ratio	Fuel consumption rate	Thrust
Zero-dimensional model	– 6.1	– 13.62	9.2	– 4.3
Multi-dimensional model	2.93	6.64	3.1	1.2

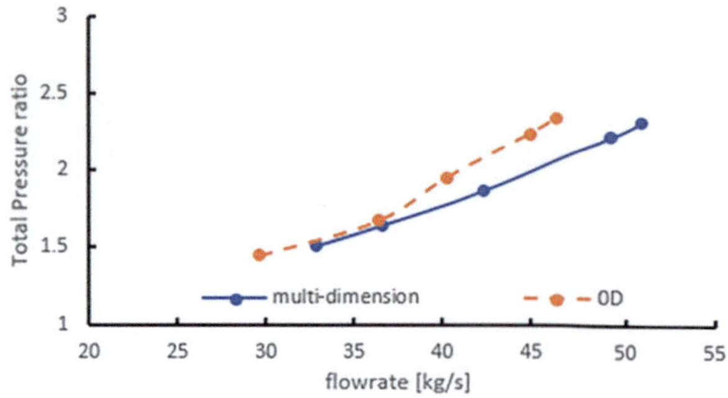

Fig. 27 Comparison of the fan total pressure ratio working lines

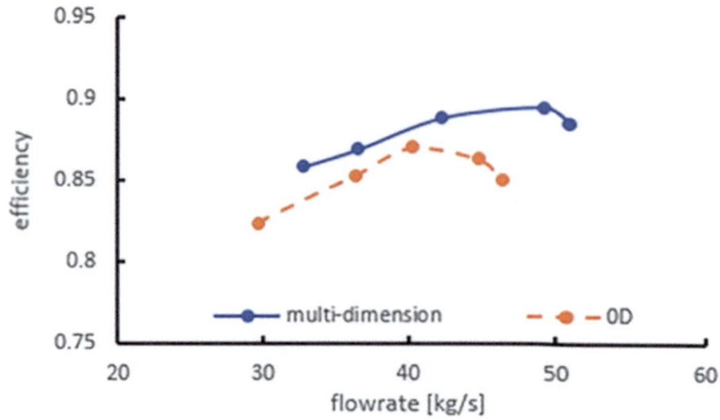

Fig. 28 Comparison of fan efficiency working lines

the overall performance parameters calculated by the multi-dimensional mixed model differ significantly from the results of the zero-dimensional model, and the calculation results are more reasonable and accurate, thus verifying the effectiveness of the turbofan zero-dimensional/three-dimensional multi-dimensional model overall performance calculation method proposed in this paper.

Conclusions

On the basis of the zero-dimensional/three-dimensional simulation model tightly coupled multi-dimensional turbofan engine overall performance calculation method proposed in this paper and the zero-dimensional model overall performance calculation program framework, an independent overall performance calculation program was developed. By combining this program with three-dimensional flow field simulation software, a multi-dimensional tightly coupled calculation framework was built, and the overall performance calculation function of the multi-dimensional model with multiple three-dimensional model components was realized. By replacing various components with three-dimensional models, the influence of the component geometry on the aerodynamic performance was incorporated to improve the accuracy of overall performance prediction.

With a mixed exhaust turbofan engine as the application object, the multi-dimensional model overall performance method was used to replace the fan components and nozzle components with three-dimensional models, and overall performance simulation results were obtained. The overall performance parameters obtained by the zero-dimensional and mixed-dimensional model calculations were compared, and the physical mechanism of the improvement in the overall performance calculation accuracy of the three-dimensional model was analyzed. By using component performance models of different dimensions for multi-dimensional simulation, the simulation accuracy of overall performance parameters, such as the engine thrust and fuel consumption rate, at the design points can be effectively improved; more accurate common working lines of fan components at off-design points can be obtained; and the balance between calculation efficiency and simulation accuracy can be flexibly adjusted according to needs.

References

1. Lytle J, Follen G, Naiman C et al (2000) Numerical propulsion system simulation (NPSS) 1999 industry review. NASA Glenn Research Center, Cleveland, Ohio, NASA/TM-2000-209795
2. Lawrence C (1997) An overview of three approaches to multidisciplinary aeropropulsion simulations. NASA Lewis Research Center, Cleveland, US, NASA TM-107443
3. Gregory F, Mira A (2000) Numerical zooming between a NPSS engine system simulation and a one-dimensional high compressor analysis code. NASA Glenn Research Center, Cleveland, US, NASA/TM-2000-209913
4. Lytle JK (2000) The numerical propulsion system simulation: an overview. NASA Glenn Research Center, Cleveland, US, NASA/TM-2000-209915
5. Yijun J (2000) Review of propulsion system numerical simulation. Gas Turbine Exp Res 13(4):56–58
6. Jie J (2003) Review of the US numerical propulsion system simulation (NPSS) program. Gas Turbine Exp Res 16(1):57–62
7. Koop WE (1997) The integrated high performance turbine engine technology (IHPTET) program. In: 13th international symposium on air breathing engines, ISABE-97-7175. Chattanooga, USA
8. Simoneau RJ, Hudson DA (1989) CFD in the context of IHPTET the integrated high performance turbine engine technology program. AIAA, Monterey, USA, AIAA-89-2904
9. Hongde J, Changde F (1989) The U.S. high performance turbine engine technology plan. Int Aviat 11:1–3
10. AIAA Air Breathing Propulsion Technical Committee (2006) The versatile affordable advanced turbine engines (VAATE) initiative. AIAA, Reston, US
11. Turner M, Ryder R, Celestina M et al (2002) High fidelity 3D turbofan engine simulation with emphasis on turboenginery-combustor coupling. In: 38th AIAA/ASME/SAE/ASEE joint propulsion conference & exhibit. AIAA, Indiana, USA, pp 1–13

12. Turner M, Norris A, Veres J (2003) High fidelity 3D simulation of the GE90 (invited). In: 33rd AIAA fluid dynamics conference and exhibit. AIAA, Orlando, USA, pp 1–13
13. Lytle J, Follen G, Naiman C et al (2002) 2001 numerical propulsion system simulation review. NASA Glenn Research Center, Cleveland, USA, NASA/TM-2002-211197
14. Medic G, You D, Kalitzin G et al (2007) Integrated computations of an entire jet engine. ASME Paper GT2007-27094
15. Shizheng Z, Genshou L (1983) Methods and computer programs for calculating the design point and off-design performance of gas turbines. J Eng Thermophys 4:321–323
16. Sellers JF, Daniele CJ (1975) Dyngen-a program for calculating steady state and transient performance of turbojet and turbofan engines. NASA Lewis Research Center, Cleveland, USA, NASA TN D-7901
17. Koenig RW, Fishbach LH (1972) GENENG-a program for calculating design and off-design performance for turbojet and turbofan engines. NASA Lewis Research Center, Cleveland, USA, NASA TN D-6552
18. Fishbach LH, Koenig RW (1972) Geneng II—a program for calculating design and off-design performance of two- and three-spool turbofans with as many as three nozzles. NASA Lewis Research Center, Cleveland, USA, NASA TN D-6553
19. Weidou N (1992) Hybrid real-time simulation of a certain type of twin-shaft aero engine. J Aerosp Power 7(4):1–9
20. Songlin L, Jianguo S, Jianmin L et al (1997) Finite field search method for solving turbofan engine mathematical model. J Aerosp Power 12(3):276–278
21. Yan Z (2000) Experimental study on the start-up acceleration process of a missile turbojet engine. Propuls Technol 21(4):5–8
22. Preece A, Bdeiwi H, Jorgens C (2021) CFD simulation of internal flow within the electric motor powered aero-engine simulator (EMPAS) assembly. AIAA 2021-2472
23. Ivanov MJ (1994) Introduction and overview. AGARD-LS-198

Nonlinear Identification Method for Aircraft Engine Component Health Parameters Based on Proximate Operating Conditions

Abstract Gas path performance analysis of aircraft engines involves the use of flight data and engine mathematical models to estimate component health parameters, which play crucial roles in increasing engine safety and reducing maintenance costs. Gas path performance analysis generally faces the problem of having many parameters to estimate (including health parameters and sensor bias errors) and few sensor measurement points. Existing health parameter estimation methods usually perform dimensionality reduction on the parameters to be estimated and cannot comprehensively and accurately estimate all the parameters. To address the issue of having few sensor measurement points as system outputs, this study uses measurement data from multiple proximate steady-state operating conditions to form an augmented system output vector, which makes the number of available system outputs greater than the number of parameters to be estimated. The homogenization of the engine model under proximate operating conditions leads to low parameter identifiability. Therefore, this study analyzes the inherent mechanism of large deviations in parameter estimation results from a linear perspective and proposes a multistage nonlinear parameter identification method that combines biased/unbiased estimation. The proposed method can accurately estimate health parameters and sensor bias errors on the basis of measurement data from proximate operating conditions. Using the JT9D engine as an example for simulation verification, the traditional gas path performance analysis method can estimate 10 health parameters with a maximum error of 2.2% (relative to the nominal values) using 10 output measurements. When 3 proximate steady-state operating conditions are used and the maximum distance between operating conditions is 4% in terms of low-pressure relative conversion speed, 23 parameters can be accurately estimated, including 10 health parameters and 13 sensor measurement bias errors, with the maximum error of the health parameters being less than 0.1%.

Keyword Aircraft engine · Gas path performance analysis · Health parameters · Nonlinear least squares

Q. Xu et al., *Digital Intelligent Aeroengine*,
https://doi.org/10.1007/978-981-95-0942-3_6

Introduction

During the long-term operation of an aircraft engine, components such as compressors and turbines may degrade due to carbon accumulation, wear, erosion, damage, and other reasons [1]. Therefore, in engine health management, gas path performance analysis, which is used to track component degradation, plays crucial roles in increasing engine safety and reducing maintenance costs. Gas path analysis (GPA) involves the use of health parameters to describe the degradation of main components such as fans, low-pressure compressors, high-pressure compressors, high-pressure turbines, and low-pressure turbines. The core strategy is to use sensor measurements of speed, temperature, and pressure along the gas path, on the basis of mathematical methods, to estimate health parameters [1]. That is, residual equations are established between sensor measurements and model calculation results on the basis of the nonlinear model at the engine component level, and a set of health parameter values are solved for when the residual equations equal zero. Traditional gas path performance analysis methods have been widely applied in practical engineering scenarios, but they still face a problem: when the number of measured parameters is less than the number of health parameters, the estimation of performance degradation becomes an underdetermined problem [2], which results in an infinite number of solutions, thus making determining the true values difficult. This situation is common because even if the number of measured parameters is not less than the number of health parameters, more parameters to be estimated are introduced because of factors such as sensor measurement bias errors, external interference, and modeling errors.

Many studies have been conducted to address this issue. Reference [3] proposed a discrete Kalman filtering method that is based on the α prior, in which the estimation of health parameters and sensor bias errors is formulated as a discrete Kalman filtering problem. The prediction results obtained using the α prior method are combined with correction information obtained using the weighted least squares method. The unique solution is determined by minimizing the norm of the parameters to be estimated. This solution is an engineering approximation of the true parameter values, and its accuracy depends on prior knowledge and external constraints. A large deviation recovery algorithm is needed to address situations where parameter changes are significant. Reference [4] used the singular value decomposition method to solve the underdetermined problem in least squares estimation, which can identify the unique solution with the smallest norm among the infinite group of solutions, which makes the changes in the estimated parameter values relative to the nominal values minimal. The estimation result has a certain degree of agreement with the sensor measurement results, and the health parameters show a reasonable change trend at multiple operating points. Reference [5] proposed an airborne adaptive model that is based on optimal regulation Kalman filtering, which reduces the dimensionality of health parameters through a transformation matrix to convert the underdetermined problem into a well-determined problem. This transformation matrix is calculated through an optimization method, and the optimization goal is to minimize the unmeasurable changes in the thrust, stability margin, etc., used for control after the health parameters are reduced. The simulation results show that although the actual health parameters cannot be obtained accurately, sufficiently accurate online estimates of engine state variables that are of concern, such as thrust, can still be obtained. Reference [6] used singular value decomposition instead of Cholesky decomposition, proposes a spherical unscented Kalman filter that can improve computational stability, and performed online estimation of health parameters; however, the number of parameters to be estimated must be equal to the number of sensor measurement outputs. Therefore, for various health parameters that cannot be estimated online, the use of transient data for offline correction is proposed. Reference [7] used prior knowledge to constrain the ranges of health parameter values and integrated constraints on state variables into the Kalman filtering algorithm by truncating the state variable probability density function. In addition, a constrained Kalman filtering method that improves the estimation accuracy for health parameters under underdetermined conditions was proposed.

The above methods cannot theoretically guarantee the accuracy of the parameter estimation results, so reference [8, 9] proposed a multioperating point analysis method that can theoretically guarantee the accuracy of the parameter estimation results. The use of sensor measurement data from multiple steady-state operating points distributed along the common working line to estimate health parameters can compensate for the problem of having an insufficient number of sensors at a single operating point. However, this method requires that the health parameters remain constant at multiple operating points, which is often not satisfied when the distance between operating points is large, which limits the practical applicability of this method. If the measurement data of nearby operating points are used to estimate the health parameters, the health parameters and the sensor bias error can be reasonably assumed to remain constant. However, the homogenization of the engine model at nearby operating points is more serious and makes the identifiability of the parameters poor.

To solve the above problems, this paper proposes a nonlinear identification method for health parameters that is based on nearby operating points. First, we modify the method of estimating health parameters using the engine steady-state nonlinear model in the traditional method to use a transient nonlinear model, thereby increasing the number of degrees of freedom for model changes to enrich the model information during parameter estimation. Then, we analyze the mathematical mechanism of low parameter identifiability when the model is homogenized at different nearby operating points from a linear perspective, and on the basis of the data collinearity analysis method, we provide a quantitative analysis method for parameter identifiability. On this basis, from a nonlinear perspective, we propose a multistage nonlinear parameter identification method that combines biased/unbiased estimation, which can effectively solve the problem of accurate estimation of health parameters via nearby operating points within a certain range.

The remainder of this article is organized as follows: Section "Mathematical Description of the Component Performance Degradation Health Parameter Estimation Problem" provides a mathematical description of the estimation problem for component health parameters and sensor bias errors, Section "Nonlinear Identification Method for Health Parameters Based on Neighboring Operating Points" introduces the multistage nonlinear identification method for component health parameters and sensor bias errors on the basis of nearby operating points, Section "Simulation Verification" uses an open-source simulation model of the JT9D engine and a set of typical component performance degradation data for simulation to verify the effectiveness of the proposed method, and Section "Conclusions" summarizes the results.

Mathematical Description of the Component Performance Degradation Health Parameter Estimation Problem

An aero engine is a complex nonlinear dynamic system, and the transient mathematical model of a twin-shaft turbofan engine can be expressed as follows:

$$\begin{cases} \dot{\mathbf{x}} = f(\mathbf{x}, \mathbf{u}) \\ \mathbf{y} = g(\mathbf{x}, \mathbf{u}) \end{cases} \tag{1}$$

where $\mathbf{x}$ is the state vector of the engine as a dynamic system. $\mathbf{x}$ has multiple selection schemes, and generally it can be expressed as $\mathbf{x} = [N_1, N_2]^T$ to represent the low-pressure and high-pressure shaft speeds. This method is used in the following text. In addition, $\mathbf{y}$ represents the system output, including the speed, temperature and pressure. $\mathbf{y}$ includes $\mathbf{x}$, and $\mathbf{y} \in R^{M_y}$, where M_y is the number of outputs. $\mathbf{u}$ denotes the engine input, which consists of two types, namely, the fuel supply, variable geometry and other control mechanisms, and the total temperature and total pressure of the intake duct (equivalent to the flight altitude and Mach number) and other external conditions; $\mathbf{u} \in R^{M_u}$, where M_u is the number of inputs. According to the structure of the engine model, from a mathematical perspective, for a steady-state operating point, the input $\mathbf{u}$ uniquely determines the operating state of the engine; for any moment in the transient process, the state quantity $\mathbf{x}$ and input $\mathbf{u}$ uniquely determine the operating state of the engine. This is important for writing residual equations in health parameter estimation.

The nonlinear function $f(\mathbf{x}, \mathbf{u})$ in Eq. (1) is generally calculated by iteratively solving a set of nonlinear algebraic equations. First, a set of initial values of the iteration variables is given. Then, the analytical models of each component are calculated in the order from front to back along the engine air path, and a set of residual equations, the number of which is equal to the number of iteration variables, is calculated. Finally, the true solution of the iteration variables that makes the residual equation equal to zero is determined iteratively by numerical methods. In the analytical models of compression components and expansion components, a set of characteristic curves for converting the flow rate and pressure ratio and a set of characteristic curves for

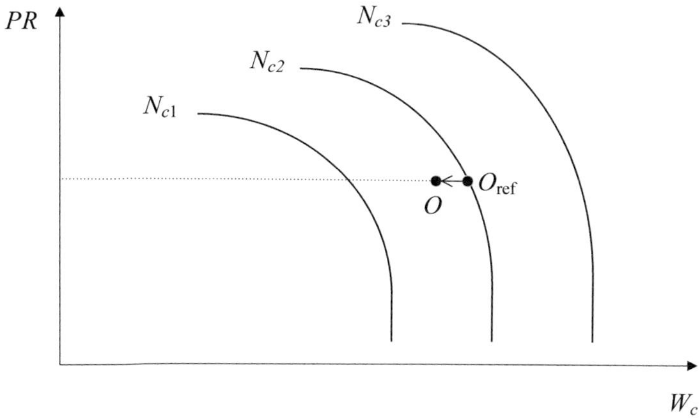

Fig. 1 Schematic diagram of the compressor characteristic curves

converting the flow rate and efficiency are needed. The characteristic curves for converting the flow rate and pressure ratio are shown in Fig. 1.

In Fig. 1, the three curves represent the corresponding relationships between the converted flow rate and pressure ratio of the compressor at different relative conversion speeds. Assuming that a certain operating point at the beginning of engine operation is at O_{ref}, when the compressor performance degrades, this operating point can be considered to move to O on the characteristic diagram, which represents a decrease in the flow rate under the same pressure ratio conditions, which is generally characterized by the flow coefficient. For a component with inlet section number i, its flow coefficient is defined as follows [10]:

$$SW_i = \frac{W_{ci}}{(W_{ci})_{ref}} = \frac{W_i\sqrt{T_{ti}/288.15}/(P_{ti}/101.325)}{\left(W_i\sqrt{T_{ti}/288.15}/(P_{ti}/101.325)\right)_{ref}} \tag{2}$$

Similarly, the component efficiency coefficient can be defined as

$$SE_i = \frac{\eta_i}{(\eta_i)_{ref}} \tag{3}$$

The initial engine flow coefficient and efficiency coefficient are both 1, which can be used as nominal values, and for convenience of calculation, the health parameters can be defined as the changes in the flow coefficient and efficiency coefficient relative to the nominal values, that is,

$$\boldsymbol{\theta}_h = [\Delta SW_i, \Delta SE_i]^T = [SW_i - 1, SE_i - 1]^T \tag{4}$$

where $\boldsymbol{\theta}_h \in R^{M_h}$. In addition, the health parameters can also include relative changes in other components, such as the nozzle throat area, duct total pressure recovery coefficient, and bleed air flow.

Owing to the many interference factors, such as measurement noise in the engine transient process, the measurement data and corresponding model of steady-state operating conditions are generally used for health parameter estimation. If the derivative of the speed in the transient model (1) is constrained to 0, the steady-state mathematical model of the engine can be obtained as follows:

$$\mathbf{y} = f_s(\mathbf{u}) \tag{5}$$

To calculate the nonlinear function f_s ($\mathbf{u}$), a set of nonlinear algebraic equations also needs to be iteratively solved, similar to the solution process for the function f ($\mathbf{x}$, $\mathbf{u}$) in Eq. (1). The main strategy for component health parameter estimation is to make the model output $\mathbf{y}$ in Eq. (5) equal to the corresponding measured value under the input condition $\mathbf{u}$, establish M_y residual equations, and then use numerical methods to solve for the health parameters that make the residual equations hold. In this process, to accurately estimate the health parameters, the sensor measurement error also needs to be considered: the zero-mean white noise error can be eliminated by taking the average of the steady-state measurement values over a period of time, whereas the bias error or constant value error cannot be eliminated, and they need to be estimated together with the health parameters as unknown parameters.

In Eq. (5), $\mathbf{y}$ and $\mathbf{u}$ are both obtained by measurement; we let $\mathbf{y}_m$ and $\mathbf{u}_m$ represent the measured values of the output and input, respectively, and let $\mathbf{y}_b$ and $\mathbf{u}_b$ represent the measurement bias errors of the output and input, respectively.

$$\mathbf{y}_m = \mathbf{y} + \mathbf{y}_b, \mathbf{u}_m = \mathbf{u} + \mathbf{u}_b \tag{6}$$

In the steady-state model (5), by introducing health parameters and measurement bias errors, the steady-state measurement output of the engine can be represented as

$$\mathbf{y}_m = \mathbf{f}_s(\mathbf{u}_m - \mathbf{u}_b, \theta_h) + \mathbf{y}_b \tag{7}$$

All the parameters to be estimated, including the health parameters and measurement bias errors, are defined as $\boldsymbol{\theta}$, as shown below:

$$\theta = [\theta_h^T, \mathbf{u}_b^T, \mathbf{y}_b^T]^T \tag{8}$$

In the formula, $\boldsymbol{\theta} \in R^{M_\theta}$, and $M_\theta = M_h + M_u + M_y$. We define $\overleftarrow{f}_s(\mathbf{u}_m, \theta) = f_s(\mathbf{u}_m - \mathbf{u}_b, \theta_h) + \mathbf{y}_b$; then, Eq. (7) can be rewritten as

$$\mathbf{y}_m = \overleftarrow{f}_s(\mathbf{u}_m, \theta) \tag{9}$$

The health parameters can be estimated by solving for the parameters θ that make Eq. (9) hold. Equation (9) shows that when the number of parameters to be estimated M_θ is greater than the number of equations M_y, the problem is underdetermined, and the true solution cannot be determined. Since θ is a small relative change from the nominal value, this problem can be further explained from a linear perspective. Equation (9) is expanded in a Taylor series at the engine target operating point ($\mathbf{u} = \mathbf{u}_m, \mathbf{y} = \mathbf{y}_m$) and the nominal value $\boldsymbol{\theta} = 0$, and only the first-order term is retained to obtain

$$\mathbf{y}_m = \overleftarrow{f}_s(\mathbf{u}_m, \boldsymbol{\theta}) = \overleftarrow{f}_s(\mathbf{u}_m, 0) + F\boldsymbol{\theta} \tag{10}$$

where $F = \frac{\partial \overleftarrow{f}_s}{\partial \boldsymbol{\theta}}(\mathbf{u}_m, 0)$, with $F \in R^{M_y \times M_\theta}$, is the partial derivative of the function $\overleftarrow{f}_s()$ with respect to $\boldsymbol{\theta}$, which includes three parts: The first is the partial derivative of the health parameter $\boldsymbol{\theta}_h$, which is affected by the engine steady-state model $f_s()$. The second is the partial derivative of the input measurement bias $\mathbf{u}_b$. A lthough it is also a sensor error, after being introduced as an input into the steady-state model (9), its effect on the model is actually very similar to that of $\boldsymbol{\theta}_h$, and its partial derivative is also affected by the steady-state model $f_s()$. The third is the partial derivative of the output measurement bias $\mathbf{y}_b$, which is equivalent to the M_y-dimensional unit matrix and is not affected by the model. The matrix F is the augmented influence coefficient matrix [2], which contains not only the conventional influence coefficient matrix that corresponds to the health parameters but also the gradient information matrix that corresponds to the measurement errors.

According to Eqs. (9) and (7), we define the nominal output $\mathbf{y}_n = \overleftarrow{f}_s(\mathbf{u}_m, 0) = f_s(\mathbf{u}_m, 0)$; then, the solution of Eq. (10) can be represented as

$$F\boldsymbol{\theta} = \mathbf{y}_m - \mathbf{y}_n \tag{11}$$

When the number of parameters M_θ is greater than the number of equations M_y, the matrix F cannot be inverted, and Eq. (11) cannot be solved.

Nonlinear Identification Method for Health Parameters Based on Neighboring Operating Points

The key problem of Eq. (9) is that the number of sensor measurements used as system outputs is not sufficient, so a natural idea is to select one or more neighboring operating points near the target operating point to supplement the measurements, thus making Eqs. (10) or (11) a well-determined or overdetermined problem so that the unique solution can be estimated by the least squares method. The premise of this method is that the health parameters and measurement bias errors change very little at several operating points, which requires the distance between different operating points to be very close; however, this easily homogenizes the engine model, which causes the model information to not be rich enough, thus making identifying the model parameters difficult. This section addresses this issue by modifying the traditional method of using the engine steady-state model for airway performance analysis to use the transient model, which makes the model information richer.

Method for Estimating Health Parameters with the Transient Model

As mentioned earlier, for the engine transient model (1), at a certain moment, the state quantity $\mathbf{x}$ and input $\mathbf{u}$ uniquely determine the working state of the engine, so the transient model (1) can be equivalent to a hypothetical engine steady-state model. With power extraction or injection on the low-pressure shaft and high-pressure shaft, the amount of power extraction or injection is determined by the speed derivative $\dot{\mathbf{x}}$ as the output of the model, whereas the rotation speed $\mathbf{x}$ is no longer considered the output of the model but is considered the input of the new equivalent model together with $\mathbf{u}$.

We define the part of the original model output $\mathbf{y}$ other than the rotation speed $\mathbf{x}$ as $\mathbf{y}^- \in R^{M_y-2}$ such that $\mathbf{y} = [\mathbf{x}^T, (\mathbf{y}^-)^T]^T$; then, the new model output can be defined as $\mathbf{z} = [\dot{\mathbf{x}}^T, (\mathbf{y}^-)^T]^T$, where $\mathbf{z} \in R^{M_y}$, and the engine transient model used for health parameter estimation can be expressed as follows:

$$\mathbf{z} = f_t(\mathbf{x}, \mathbf{u}) \tag{12}$$

Notably, although we use a transient model, the measurement data used still come from steady-state operating points, so the measurement value of $\dot{\mathbf{x}}$ must be 0, there is no bias error, and the parameter to be identified is still $\boldsymbol{\theta}$.

Compared with the steady-state model (5), Eq. (12) uses $\mathbf{x}$ as input, so the degrees of freedom of model changes can be relaxed when calculating the partial derivatives of the parameters to be identified (because $\dot{\mathbf{x}}$ need not be zero), thereby obtaining more gradient information from the model, which is very beneficial for improving the identifiability of the parameters. We verify this through simulation in Section "Simulation Verification".

Linear Least Squares Method for Health Parameter Estimation Based on Neighboring Operating Points

N nearby operating points are selected, and the input and output measurement values of each operating point are $(\mathbf{u}_m^i, \mathbf{x}_m^i, \mathbf{z}_m^i)$, where $i = 1, \ldots, N$. Each measurement value is the average over multiple actual sampling points.

First, the following assumptions are made:

(1) At N nearby operating points, the degrees of change in the health parameters are very small, and these parameters can be considered constant;
(2) At N nearby operating points, the change in the sensor measurement bias error is very small, and the sensor measurement bias error can be considered constant.

We consider N operating points and introduce health parameters and sensor bias errors into Eq. (13) to obtain

$$\left\{\mathbf{z}_m^i = f_t\left(\mathbf{x}_m^i - \mathbf{x}_b, \mathbf{u}_m^i - \mathbf{u}_b, \boldsymbol{\theta}_h\right) + \mathbf{v}^i + \mathbf{z}_b\,, i = 1, 2, \ldots, N\right. \tag{13}$$

where $\mathbf{v}^i$ is the model error caused by various modeling mechanisms and external disturbances not considered in f_t () and $\mathbf{z}_b$ is the measurement bias error of output $\mathbf{z}$ in Eq. (12). We define $\overleftarrow{f}_t\left(\mathbf{x}_m^i, \mathbf{u}_m^i, \boldsymbol{\theta}\right) = f_t\left(\mathbf{x}_m^i - \mathbf{x}_b, \mathbf{u}_m^i - \mathbf{u}_b, \boldsymbol{\theta}_h\right) + \mathbf{z}_b$; then, Eq. (13) can be rewritten as follows:

$$\left\{\mathbf{z}_m^i = \overleftarrow{f}_t\left(\mathbf{x}_m^i, \mathbf{u}_m^i, \boldsymbol{\theta}\right) + \mathbf{v}^i\,, i = 1, 2, \ldots, N\right. \tag{14}$$

where $\boldsymbol{\theta} = \left[\boldsymbol{\theta}_h^T, \mathbf{u}_b^T, \mathbf{x}_b^T, \mathbf{y}'^T_b\right]^T = \left[\boldsymbol{\theta}_h^T, \mathbf{u}_b^T, \mathbf{y}_b^T\right]^T$. According to the measurement values of N operating points $\left(\mathbf{u}_m^i, \mathbf{x}_m^i, \mathbf{z}_m^i\right)$ and the equation group in Eq. (14), the parameter $\boldsymbol{\theta}$ can be solved for. If enough nearby operating points are selected to satisfy $M_{\overleftarrow{z}} = N \times M_y \geq M_\theta$, then solving Eq. (14) can become a well-defined or overdetermined problem, and the least squares solution with the smallest sum and variance can be calculated [11]. We first use linear least squares estimation for parameter identification to provide a clear explanation of the essence of the problem, which also corresponds to the commonly used linear airway performance analysis method [12]. The combined parameter $\boldsymbol{\theta}$ can be considered prior knowledge of small perturbations, and Eq. (14) can be expanded by Taylor expansion to obtain

$$\left\{\mathbf{z}_m^i = \overleftarrow{f}_t\left(\mathbf{x}_m^i, \mathbf{u}_m^i, 0\right) + \frac{\partial \overleftarrow{f}_t}{\partial \boldsymbol{\theta}}\left(\mathbf{x}_m^i, \mathbf{u}_m^i, 0\right)\boldsymbol{\theta} + \mathbf{v}^i\,, i = 1, 2, \ldots, N\right. \tag{15}$$

In Eq. (15), the model error $\mathbf{v}^i$ also includes the high-order terms after small perturbation linearization. Equation (15) can be rewritten as

$$\overleftarrow{\mathbf{z}}_e = \overleftarrow{F}\boldsymbol{\theta} + \overleftarrow{\mathbf{v}} \tag{16}$$

$$\overleftarrow{\mathbf{z}}_e = \begin{bmatrix} \mathbf{z}_m^1 - \overleftarrow{f}_t(\mathbf{x}_m^1, \mathbf{u}_m^1, 0) \\ \mathbf{z}_m^2 - \overleftarrow{f}_t(\mathbf{x}_m^2, \mathbf{u}_m^2 0) \\ \vdots \\ \mathbf{z}_m^N - \overleftarrow{f}_t(\mathbf{x}_m^N, \mathbf{u}_m^N, 0) \end{bmatrix} \tag{17}$$

$$\overleftarrow{F} = \begin{bmatrix} \frac{\partial \overleftarrow{f}_t}{\partial \boldsymbol{\theta}}(\mathbf{x}_m^1, \mathbf{u}_m^1, 0) \\ \frac{\partial \overleftarrow{f}_t}{\partial \boldsymbol{\theta}}(\mathbf{x}_m^2, \mathbf{u}_m^2, 0) \\ \vdots \\ \frac{\partial \overleftarrow{f}_t}{\partial \boldsymbol{\theta}}(\mathbf{x}_m^N, \mathbf{u}_m^N, 0) \end{bmatrix} \tag{18}$$

$$\overleftarrow{\mathbf{v}} = \left[\mathbf{v}^1 \ldots \mathbf{v}^N\right]^T \tag{19}$$

In the formula, the matrix $\overleftarrow{F}$ is a combination of the Jacobian matrices calculated by the function $\overleftarrow{f}()$ at the N operating points. Then, in the least squares sense, the optimal estimate of $\boldsymbol{\theta}$ is obtained by minimizing the sum of the squares of the errors between the measured values and the model output [11]:

$$J(\boldsymbol{\theta}) = \frac{1}{2}\left(\overleftarrow{\mathbf{z}}_e - \overleftarrow{F}\boldsymbol{\theta}\right)^T\left(\overleftarrow{\mathbf{z}}_e - \overleftarrow{F}\boldsymbol{\theta}\right) \tag{20}$$

According to the linear least squares method, the parameter estimate that minimizes the cost function $J(\boldsymbol{\theta})$ is $\hat{\theta}$:

$$\hat{\boldsymbol{\theta}} = \left(\overleftarrow{F}^T \overleftarrow{F}\right)^{-1} \overleftarrow{F}^T \overleftarrow{\mathbf{z}}_e \tag{21}$$

Analysis of Data Collinearity via Linear Least Squares Estimation

The identifiability of the parameter $\boldsymbol{\theta}$ is analyzed from the perspective of the linear model. This paper considers only the parameter estimation problem under the determined model structure (i.e., the selection method of the parameters to be identified is determined), so the model structure can be assumed to be accurate. Therefore, in the least squares estimation problem (16), the identifiability of $\boldsymbol{\theta}$ is determined mainly by the degree of data collinearity of the matrix $\overleftarrow{F}$ [11]. We first provide an intuitive explanation of data collinearity and then analyze the adverse effects of data collinearity on parameter identification.

In Eq. (18), the matrix $\overleftarrow{F}$ is composed of the partial derivative matrices at the N neighboring operating points $\overleftarrow{F}^i = \frac{\partial \overleftarrow{f}_t}{\partial \boldsymbol{\theta}}\left(\mathbf{x}_m^i, \mathbf{u}_m^i, 0\right)$. In extreme cases, if multiple neighboring operating points are chosen as the same point, the $\overleftarrow{F}^i$ values of the different operating points are equal and are completely linearly related in algebraic terms; then, the rank of the matrix $\overleftarrow{F}$ is equal to the rank of $\overleftarrow{F}^i$. For the underdetermined problem at a single point, $\overleftarrow{F}^i$ is not full rank, so $\overleftarrow{F}$ is also not full rank; then, we have [11]

$$\mathrm{Det}\left(\overleftarrow{F}^T \overleftarrow{F}\right) = 0 \tag{22}$$

In addition, $\left(\overleftarrow{F}^T \overleftarrow{F}\right)^{-1}$ does not exist. When the partial derivative matrices $\overleftarrow{F}^i$ are very close to each other, their degree of linear correlation is very strong, and Eq. (22) is approximately true, which indicates that the degree of data collinearity of the matrix $\overleftarrow{F}$ is very strong, and the matrix $\left(\overleftarrow{F}^T \overleftarrow{F}\right)$ is called ill-conditioned. Because the matrix $\left(\overleftarrow{F}^T \overleftarrow{F}\right)^{-1}$ is used in linear least squares estimation (21), this will cause a singular calculation problem and reduce the accuracy of parameter estimation. The identifiability of the parameters at this time can be considered poor.

The degree of data collinearity can be analyzed quantitatively by the condition number of the matrix $\overleftarrow{F}$ [13]. The singular value decomposition of the matrix $\overleftarrow{F}$ can be obtained as follows:

$$SVD\left(\overleftarrow{F}\right) = USV^T \tag{23}$$

In the formula, $U \in R^{M_{\overleftarrow{z}} \times M_\theta}$ and $V \in R^{M_\theta \times M_\theta}$ are both orthogonal matrices, that is, they satisfy $U^T U = V^T V = I$, and $S \in R^{M_\theta \times M_\theta}$ is a diagonal matrix whose the diagonal elements $\mu_j (j = 1, 2, ..., N_\theta)$ are the singular values of the matrix $\overleftarrow{F}$. The condition number of the matrix $\overleftarrow{F}$ is defined as the ratio of the maximum singular value $\mu_{\max}$ to the minimum singular value $\mu_{\min}$:

$$\mathrm{Cond}\left(\overleftarrow{F}\right) = \frac{\mu_{\max}}{\mu_{\min}} \tag{24}$$

The larger the condition number of the matrix $\overleftarrow{F}$ is, the greater the degree of data collinearity. The ranges of values of different actual physical variables are different, so comparing the condition numbers of different types of matrices is difficult. A feasible method in engineering is to normalize the columns of the matrix $\overleftarrow{F}$ [13] to unify the comparison standard.

Strong data collinearity tends to produce large absolute values of least squares estimates. The specific impact is analyzed below. According to Eq. (16), the true value of the parameter $\boldsymbol{\theta}$ can be regarded as

$$\boldsymbol{\theta} = \left(\overleftarrow{F}^T \overleftarrow{F}\right)^{-1} \overleftarrow{F}^T \left(\overleftarrow{\mathbf{z}}_e - \overleftarrow{\mathbf{v}}\right) \tag{25}$$

For ease of analysis, $\overleftarrow{\mathbf{v}}$ can be assumed to represent Gaussian white noise with uncorrelated members [11], namely,

$$E\left(\overleftarrow{\mathbf{v}}\right) = 0E\left(\overleftarrow{\mathbf{v}}\,\overleftarrow{\mathbf{v}}^T\right) = \sigma^2 I \tag{26}$$

Then, combined with least squares estimation (21), the covariance matrix of the parameter estimation error $\hat{\theta} - \boldsymbol{\theta}$ can be obtained [11]:

$$\begin{aligned} \mathrm{Cov}\left(\hat{\boldsymbol{\theta}}\right) &\equiv E\left[\left(\hat{\boldsymbol{\theta}} - \boldsymbol{\theta}\right)\left(\hat{\boldsymbol{\theta}} - \boldsymbol{\theta}\right)^T\right] \\ &= \left(\overleftarrow{F}^T \overleftarrow{F}\right)^{-1} \overleftarrow{F}^T E\left(\overleftarrow{\mathbf{v}}\overleftarrow{\mathbf{v}}^T\right) \overleftarrow{F} \left(\overleftarrow{F}^T \overleftarrow{F}\right)^{-1} \\ &= \sigma^2 \left(\overleftarrow{F}^T \overleftarrow{F}\right)^{-1} \end{aligned} \tag{27}$$

By substituting the singular value decomposition (27) of the matrix $\overleftarrow{F}$, we obtain

$$\mathrm{Cov}(\hat{\theta}) = \sigma^2 \Lambda^{-1} \tag{28}$$

where Λ is a diagonal matrix whose diagonal elements are the squares of the singular values of the matrix $\overleftarrow{F}$, that is, $\Lambda = \mathrm{Diag}\left(\mu_1^2, \mu_2^2, \ldots \mu_{M_\theta}^2\right)$. The parameter estimate $\hat{\theta}$ without model structure misspecification bias can be expressed in terms of the mean square error [11]:

$$\begin{aligned} \mathrm{MSE}(\hat{\boldsymbol{\theta}}) &\equiv E\left[(\hat{\boldsymbol{\theta}} - \boldsymbol{\theta})^T (\hat{\boldsymbol{\theta}} - \boldsymbol{\theta})\right] = \sum_{j=1}^{M_\theta} E(\hat{\theta}_j - \theta_j)^2 \\ &= \sigma^2 \mathrm{Tr}\left[\left(\overleftarrow{F}^T \overleftarrow{F}\right)^{-1}\right] = \sigma^2 \sum_{j=1}^{M_\theta} \mu_j^{-2} \end{aligned} \tag{29}$$

According to Eq. (29), when the matrix $\overleftarrow{F}$ contains smaller singular values, the amplitude of the model error term $\mathbf{v}$ in the parameter estimation error is greatly amplified, which causes the least squares estimation result to deviate significantly from the true value.

For general least squares estimation engineering applications, when the number of conditions of the correlation matrix is greater than 30 after normalization, the data collinearity is strong, and the identification result accuracy is low. Strong data collinearity essentially indicates that the information contained in the data is not rich enough. For the linear identification problem for the engine parameters $\boldsymbol{\theta}$ in Eq. (16), since the matrix $\overleftarrow{F}$ contains model information of adjacent operating points, the condition number is generally large, which results in low accuracy of the linear method identification results.

To address this issue, we propose a multistage nonlinear least squares estimation algorithm that is based on a mixture of biased/unbiased estimates to perform nonlinear identification of the parameters $\boldsymbol{\theta}$ in Eq. (14). The nonlinear least squares method is an iterative algorithm that uses gradient information to update the parameter estimate $\hat{\theta}$ from a linear perspective at each step; the most common such algorithm is the Newton–Raphson method [11, 14]. Therefore, in each step of iteration, if the data collinearity of the matrix $\overleftarrow{F}$ is strong, then $\hat{\theta}$ will gradually diverge, and the use of biased estimates can improve this situation to make $\hat{\theta}$ converge.

Multistage Nonlinear Least Squares Estimation Algorithm Based on a Mixture of Biased/Unbiased Estimates

The nonlinear parameter estimation problem in Eq. (14) can be rewritten in the form of the following nonlinear least squares identification model:

$$\overleftarrow{\mathbf{z}}_m = \overleftarrow{f}_t\left(\overleftarrow{\mathbf{x}}_m, \overleftarrow{\mathbf{u}}_m, \boldsymbol{\theta}\right) + \overleftarrow{\mathbf{v}} \tag{30}$$

$$\overleftarrow{\mathbf{z}}_m = \left[\mathbf{z}_m^1 \ldots \mathbf{z}_m^N\right]^T, \overleftarrow{\mathbf{x}}_m = \left[\mathbf{x}_m^1 \ldots \mathbf{x}_m^N\right]^T, \overleftarrow{\mathbf{u}}_m = \left[\mathbf{u}_m^1 \ldots \mathbf{u}_m^N\right]^T \tag{31}$$

The parameter estimate can be obtained by minimizing the following cost function [11]:

$$J(\boldsymbol{\theta}) = \frac{1}{2}[\overleftarrow{\mathbf{z}}_m - \overleftarrow{f}_t\left(\overleftarrow{\mathbf{x}}_m, \overleftarrow{\mathbf{u}}_m, \boldsymbol{\theta}\right)]^T \left[\overleftarrow{\mathbf{z}}_m - \overleftarrow{f}_t\left(\overleftarrow{\mathbf{x}}_m, \overleftarrow{\mathbf{u}}_m, \boldsymbol{\theta}\right)\right] \tag{32}$$

To solve the problem, nonlinear iterative optimization techniques must be used; here, we directly present the commonly used Newton–Raphson iterative solution algorithm [11, 14]:

$$\hat{\boldsymbol{\theta}}_{k+1} = \hat{\boldsymbol{\theta}}_k + \left(\overleftarrow{F}_k^T \overleftarrow{F}_k\right)^{-1} \overleftarrow{F}_k^T \left[\overleftarrow{\mathbf{z}}_m - \overleftarrow{F}_t\left(\overleftarrow{\mathbf{x}}_m, \overleftarrow{\mathbf{u}}_m, \hat{\boldsymbol{\theta}}_k\right)\right] \tag{33}$$

$$\overleftarrow{F}_k = \frac{\partial \overleftarrow{f}_t}{\partial \boldsymbol{\theta}}\left(\overleftarrow{\mathbf{x}}_m, \overleftarrow{\mathbf{u}}_m, \hat{\theta}_k\right) = \begin{bmatrix} \frac{\partial \overleftarrow{f}_t}{\partial \boldsymbol{\theta}}\left(\mathbf{x}_m^1, \mathbf{u}_m^1, \hat{\theta}_k\right) \\ \frac{\partial \overleftarrow{f}_t}{\partial \boldsymbol{\theta}}\left(\mathbf{x}_m^2, \mathbf{u}_m^2, \hat{\theta}_k\right) \\ \vdots \\ \frac{\partial \overleftarrow{f}_t}{\partial \boldsymbol{\theta}}\left(\mathbf{x}_m^N, \mathbf{u}_m^N, \hat{\theta}_k\right) \end{bmatrix} \tag{34}$$

The matrix $\overleftarrow{F}_k$ is a combination of the Jacobian matrices of the function $\overleftarrow{f}_t()$ at the N operating points at $\boldsymbol{\theta} = \hat{\theta}_k$. When Eqs. (21) and (33) are compared, nonlinear least squares estimation using the Newton–Raphson method is actually a process of multiple iterations of linear least squares estimation, and the parameters to be estimated need to be updated and the matrix $\overleftarrow{F}_k$ needs to be recalculated after each iteration.

The analysis of the linear least squares problem in the previous section still applies. If the condition number of the matrix $\overleftarrow{F}_k$ is large in the kth iteration, then the parameter estimation bias in this step will also be large. The difference is that in nonlinear least squares estimation, subsequent iterations correct this bias, so the tolerance for the degree of data collinearity of the matrix $\overleftarrow{F}_k$ is much greater. However, when the condition number of the matrix $\overleftarrow{F}_k$ is relatively large, Eq. (29) and a related analysis show that the parameter $\hat{\theta}_k$ greatly amplifies the interference term $\overleftarrow{\mathbf{v}}$ during the update, which causes the parameter estimate to deviate too much from the true value and diverge. To solve this problem, we propose a multistage nonlinear least squares estimation algorithm that is based on a mixture of biased/unbiased estimates.

The ordinary least squares method theoretically provides linear unbiased estimation results. Another identification method is biased estimation. In biased estimation, theoretically, the results always have bias, but biased estimation may produce results with smaller mean square errors [11]. Therefore, when the data are ill-conditioned, the bias produced by biased estimation may be much smaller than that produced by ordinary least squares. An easy-to-implement biased estimation method is the rank deficiency method [11, 15], which involves calculating the reduced-rank inverse of the matrix $\overleftarrow{F}_k^T \overleftarrow{F}_k$ in Eq. (33) on the basis of singular value decomposition.

We suppose that the singular value decomposition of the matrix $\overleftarrow{F}_k^T \overleftarrow{F}_k$ is

$$SVD\left(\overleftarrow{F}_k^T \overleftarrow{F}_k\right) = U_k S_k V_k^T \tag{35}$$

and we suppose that its singular values are arranged from large to small as follows:

$$\lambda_1 > \lambda_2 > \cdots > \lambda_{M_\theta - m} > \lambda_{M_\theta - m+1} > \cdots > \lambda_{M_\theta} \tag{36}$$

By deleting the smallest m singular values in Eq. (36), the reduced-rank inverse can be obtained as follows:

$$\underset{inv\_k}{\overleftarrow{F}} = V_k D_k U_k^T \tag{37}$$

$$D_k = \mathrm{Diag}\left(\left[\,1/\lambda_1 \;\; 1/\lambda_2 \;\cdots\; 1/\lambda_{M_\theta - m} \;0 \cdots 0\right]^T\right) \tag{38}$$

According to the analysis of Eqs. (25–29) in Section "Analysis of Data Collinearity via Linear Least Squares Estimation", when the smallest m singular values are deleted, the mean square error of the parameter estimate calculated using the reduced-rank inverse F_{inv_k} is greatly reduced. The iterative solution algorithm is as follows:

$$\hat{\boldsymbol{\theta}}_{k+1} = \hat{\boldsymbol{\theta}}_k + \overleftarrow{F}_{inv\_k} \overleftarrow{F}_k^T \left[\overleftarrow{\mathbf{z}}_m - \overleftarrow{f}_t\left(\overleftarrow{\mathbf{x}}_m, \overleftarrow{\mathbf{u}}_m, \hat{\boldsymbol{\theta}}_k\right)\right] \tag{39}$$

The value of m can be determined according to the singular values in specific applications.

The parameter estimate obtained by simply using the above biased estimation method is biased, so in the initial stage, Eq. (39) can be iterated several times to obtain a parameter estimate that is close to the true value, and then the unbiased estimation method can be used to continue iterating on this basis. The specific steps are as follows:

(1) In the first stage, with $\hat{\theta} = 0$ as the initial value, m_1 singular values are removed, and the rank deficiency method and Eq. (39) are used to iterate until the convergent solution $\hat{\theta} = \boldsymbol{\theta}^1$ is obtained;

……

(i) In stage i, with $\hat{\theta} = \boldsymbol{\theta}^{i-1}$ as the initial value, m_i singular values are removed, and the rank deficiency method and Eq. (39) are used to iterate until the convergent solution $\hat{\theta} = \boldsymbol{\theta}^i$ is obtained;

……

(n) With $\hat{\theta} = \boldsymbol{\theta}^{n-1}$ as the initial value, Eq. (33) is used to iterate until the final convergent solution $\hat{\theta} = \boldsymbol{\theta}^n$ is obtained.

In the last step, all singular values of $\overleftarrow{F}_k^T \overleftarrow{F}_k$ are retained, and the theoretical true value of the parameter can be obtained using the unbiased estimation method. Of course, in practical engineering applications, if the model error term $\mathbf{v}$ is too large or the degree of collinearity of the matrix is too high, thus causing the unbiased estimation to not converge, biased estimation can still be used to obtain an approximate solution. The above multistage least squares estimation algorithm can effectively improve convergence when health parameters are estimated on the basis of neighboring operating points, which is verified in the next section through a specific engine model simulation.

Simulation Verification

Introduction to the Engine Object

This paper uses a civil turbofan engine JT9D [16, 17] as the object for the simulation verification of the proposed method; its configuration and section numbers are shown in Fig. 2.

The engine is a high-bypass-ratio twin-spool turbofan engine, as shown in Fig. 2, with the main components including a fan, low-pressure rotor, gearbox, bypass duct, low-pressure compressor, high-pressure compressor, high-pressure rotor, combustion chamber, high-pressure turbine, low-pressure turbine, main duct tailpipe, and bypass duct tailpipe. According to references [16, 17], the JT9D engine can use the sensors shown in Table 1 for gas path performance analysis, and we also provide a set of typical measurement bias errors for subsequent simulation verification.

In accordance with the transient model (12) used for parameter estimation, sensors 1–5# in Table 1 are used as model inputs to determine the working conditions of the engine, and sensors 6–13# together with two speed derivatives, serve as model outputs, that is, **z** in Eq. (12).

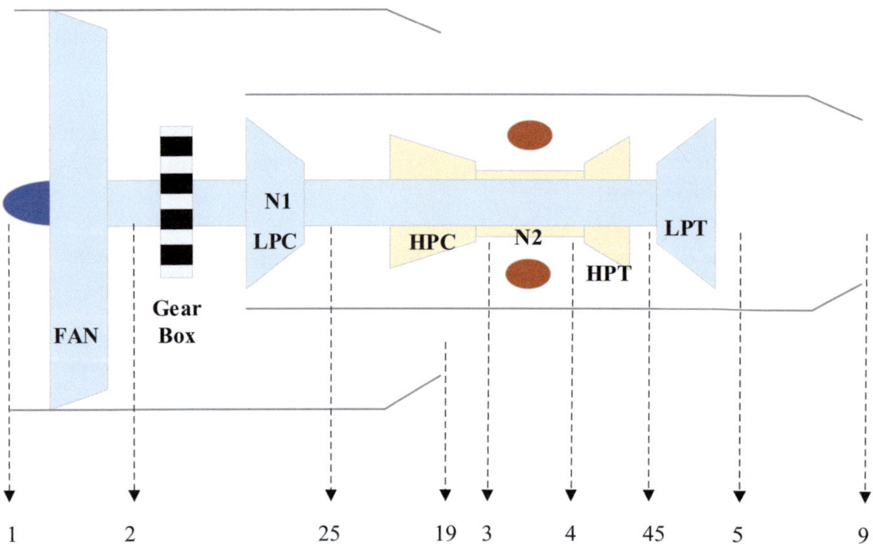

Fig. 2 Schematic diagram of the JT9D engine configuration

Table 1 Sensors for gas path performance analysis and a set of typical bias errors

Serial number	Sensor	Unit	Bias error
1	P_{t1}	kPa	0.1
2	T_{t1}	K	1
3	W_f	kg/s	− 0.004536
4	N_1	RPM	− 2
5	N_2	RPM	− 3
6	P_{s25}	kPa	0.1
7	P_{t25}	kPa	0.1
8	T_{t25}	K	2
9	P_{s3}	kPa	0.1
10	T_{t3}	K	− 2
11	T_{t45}	K	− 5
12	P_{t5}	kPa	0.2
13	T_{t5}	K	5

The health parameters that represent the degradation of component airway characteristics include the flow and efficiency correction coefficients of 5 compression/expansion components, which are defined in Eqs. (2) and (3). References [16–18] provide a set of typical health parameter values for the JT9D engine at a moderate level of component degradation, as shown in Table 2.

When degradation occurs, the flow and efficiency coefficients of most components decrease relative to the nominal values, and only the flow coefficients of two turbines increase.

The 10 health parameters in Table 2 and the 13 sensor bias errors in Table 1 form a total of 23 parameters to be identified as $\boldsymbol{\theta}$, whereas only 10 sensor measurements can be used as model outputs (which means that only 10 residual equations can be listed at a single operating point). We used the publicly available component-level simulation model of the JT9D engine (from the NASA open-source MATLAB toolbox T-MATS) [19] for simulation verification. First, we introduced health parameters and sensor bias errors into the simulation model. We then used the health parameter nonlinear identification method that is based on neighboring operating points for parameter estimation, and we compared the results with those of the traditional method based on a single operating point.

Table 2 JT9D health parameters

Corresponding component	Fan		LPC		HPC		HPT		LPT	
Health parameter (%)	ΔSW_1	ΔSE_1	ΔSW_2	ΔSE_2	ΔSW_{25}	ΔSE_{25}	ΔSW_4	ΔSE_4	ΔSW_{45}	ΔSE_{45}
Value	− 2.04	− 1.5	− 2.08	− 1.46	− 3.91	− 2.94	1.76	− 2.63	0.25	− 0.54

Simulation Results and Analysis

When neighboring operating points are used for health parameter estimation, determining the distance between operating points is crucial: if the distance is too far, the assumption that health parameters remain constant may no longer hold; if the distance is too close, the homogenization of models at different operating points is severe, which causes severe data collinearity, and obtaining accurate estimates of the health parameters is impossible. Therefore, the distance between neighboring operating points should be as close as possible until the severity of data collinearity prevents accurate estimates of health parameters from being obtained. The closest distance is determined by simulating the distances between different operating points. Since the relationships between the component characteristics and the conversion speed are relatively close, we use the difference in the low-pressure conversion speed to measure the distance between different operating points. For the JT9D engine, theoretically, accurate estimates of the 23 parameters to be identified can be obtained using 30 sensor measurements from 3 neighboring operating points, and this method was used in subsequent simulations.

To compare and verify the advantages of the proposed method, we first simulated the traditional method based on a single operating point, with 10 sensor measurements used to estimate 10 health parameters, and did not identify sensor measurement errors. Then, the health parameter identification method that is based on 3 neighboring operating points was simulated, and four sets of operating point combinations were selected, which represented different maximum distances of low-pressure conversion speed, as shown in Table 3. Moreover, for the first combination, two different methods that use the steady-state model (5) and the transient model (12) were compared. The neighboring operating point method in Table 3, unless otherwise stated, refers to the method that uses the transient model (12). The initial condition number listed in Table 3 is the condition number of the model-related part $\overleftarrow{F}$ in the matrix $\overleftarrow{F}_{m}$ when each parameter to be identified takes an initial value of zero; the specific meaning will be explained later.

When the multistage nonlinear identification algorithm is used, determining the number of stages and the number of singular values discarded at each stage is also crucial, and they can be determined by trial and error in the simulation process. Eventually, we used a three-stage nonlinear identification algorithm in the simulation, and the numbers of singular values discarded at each stage were $m = 9$, $m = 1$, and $m = 0$. The identification results of the two methods for the health parameters are shown in Table 4, and the identification results for the sensor bias errors are shown in Table 5.

(1) Traditional single-point method

For the single-point method, the condition number in Table 3 is the condition number of the impact factor matrix $\overleftarrow{F}$, and its value is very large, indicating that the degree of data collinearity is very high, which is due mainly to P_{s25} and P_{t25}. The collinearity

Table 3 Settings of two health parameter estimation methods for JT9D

Health parameter estimation method		Number of operating points	Low-pressure conversion speed N_{IC} (%)	N_{IC} maximum distance (%)	Initial condition number
Traditional single-point method		1	91.58	–	$5e^{16}$
Nonlinear identification method that is based on neighboring operating points	Operating point combination 1 (steady-state model)	3	88.79, 91.58, 94.34	5.55	863
	Operating point combination 1 (transient model)	3	88.79, 91.58, 94.34	5.55	2093
	Operating point combination 2	3	89.6, 91.98, 94.34	4.74	2310
	Operating point combination 3	3	90.39, 92.35, 94.34	3.95	2989
	Operating point combination 4	3	91.18, 92.73, 94.34	3.16	3657

of these two sensors is very high. At this point, the matrix is close to singular, so the Newton–Raphson method cannot be used directly to solve for the parameters to be identified. After the rank-deficient method is used to discard the 2 smallest singular values, the results in Table 4 are obtained. By using the traditional single-point method, high-precision estimation results can be obtained for several health parameters, such as the fan efficiency coefficient ΔSE_2, the high-pressure turbine flow coefficient ΔSW_4 and the high-pressure turbine efficiency coefficient ΔSE_4, but the estimation accuracy for the other health parameters is lower because the sensor bias error is not considered.

(2) Adjacent operating point method—Combination 1

For operating point combination 1 with a maximum distance of 5.55%, the steady-state model and transient model were used for parameter identification. As shown in Table 3, the condition number of the steady-state model is smaller than that of the transient model because when the condition number is calculated, only the column vectors related to the model in the matrix $\overleftarrow{F}$ are used, and the number of column vectors that correspond to the steady-state model is less than the number of column vectors that correspond to the transient model. If the condition number of the transient model is calculated using the same column vectors as the steady-state model is, the condition number is 764, which is less than the condition number of the steady-state model in Table 3. As shown in Tables 4 and 5, the estimation accuracy of the steady-state model is much lower than that of the transient model. According to Table 4, compared with the single-point method using the steady-state model, the adjacent operating point method has greatly improved the estimation accuracy of health parameters, and the results of the adjacent operating point method with

Table 4 JT9D health parameter identification results

Parameter name	True value (%)	Single-point method (%)	Neighboring operating point method (%)				
			Combination 1 (steady-state model)	Combination 1 (transient model)	Combination 2	Combination 3	Combination 4
ΔSW_1	− 2.04	− 2.7806	− 2.04	− 2.0392	− 2.0351	− 2.04	− 1.4925
ΔSE_1	− 1.5	− 0.39748	− 1.0524	− 1.5066	− 1.477	− 1.4998	− 0.0461
ΔSW_2	− 2.08	− 0.0995	− 2.3136	− 2.0722	− 2.1018	− 2.08	− 5.4251
ΔSE_2	− 1.46	− 1.5516	− 1.6654	− 1.4535	− 1.4864	− 1.46	− 5.6207
ΔSW_{25}	− 3.91	− 2.6939	− 4.3085	− 3.9494	− 3.9213	− 3.9113	0.4624
ΔSE_{25}	− 2.94	− 1.501	− 3.9412	− 2.9395	− 2.9365	− 2.9397	− 2.5794
ΔSW_4	1.76	1.7033	1.8034	1.7609	1.7602	1.76	1.6444
ΔSE_4	− 2.63	− 2.4291	− 0.3984	− 2.5999	− 2.629	− 2.6298	− 5.6286
ΔSW_{45}	0.25	− 0.9271	− 1.2094	0.2266	0.2501	0.2499	1.7896
ΔSE_{45}	− 0.54	− 2.7862	− 1.2449	− 0.553	− 0.5399	− 0.54	0.4571
Variance	–	$1.6e^{-3}$	$9e^{-4}$	$3.3e^{-7}$	$1.9e^{-7}$	$1.8e^{-10}$	$6.3e^{-3}$

Table 5 JT9D sensor bias error identification results

Parameter name	True value	Neighboring operating point method				
		Combination 1 (steady-state model)	Combination 1 (transient model)	Combination 2	Combination 3	Combination 4
ΔP_{t1} (kPa)	0.1	– 0.0322	0.1031	0.0952	0.0999	– 0.0942
ΔT_{t1} (K)	1	0.0928	1.0141	0.99063	0.9993	4.6291
ΔW_f (kg/s)	– 0.0045	– 0.0045	– 0.0046	– 0.0045	– 0.0045	0.0003
ΔN_1 (RPM)	– 2	– 6.3453	– 1.9559	– 2.0162	– 2.0049	41.502
ΔN_2 (RPM)	– 3	– 19.506	– 3.725	– 3.355	– 3.0359	131.39
ΔP_{s25} (kPa)	0.1	– 0.2617	0.1097	0.0961	0.0998	2.283
ΔP_{t25} (kPa)	0.1	– 0.2591	0.1093	0.0959	0.0998	2.3423
ΔT_{t25} (K)	2	0.7751	2.0223	1.9889	1.9992	7.421
ΔP_{s3} (kPa)	0.1	– 0.1716	0.063716	0.077	0.0985	9.0183
ΔT_{t3} (K)	– 2	– 6.7367	– 2.0094	– 2.017	– 2.0015	9.6977
ΔT_{t45} (K)	– 5	– 6.5534	– 4.9684	– 5.0027	– 5.0005	– 2.1474
ΔP_{t5} (kPa)	0.2	0.2004	0.1977	0.1984	0.1999	0.8055
ΔT_{t5} (K)	5	3.6974	5.0155	4.9984	4.9997	7.1742
Variance	–	320.6	0.5	0.1	$1.3e^{-3}$	$2e^{4}$

the transient model can match the true values. As shown in Table 5, the estimation accuracy for the sensor bias error obtained by the adjacent operating point method with the steady-state model is low, whereas the results of the transient model are more accurate. When health parameters are estimated, although the measurement data come from the steady-state process, the identification accuracy can be greatly improved after the transient model is modified.

For the adjacent operating point method with the transient model, Tables 6 and 7 present the identification results at the three stages of $m = 9$, $m = 1$, and $m = 0$ and the results of direct unbiased estimation without the use of the multistage method.

For the single-stage unbiased estimation method, because the initial matrix condition number is large, the model error is greatly amplified, so the iteration process exhibits numerical divergence; the results of the first step of iteration are presented in Tables 6 and 7, which show that there is already a large deviation at this time.

For the multistage method, in the first stage, biased estimation is first used to obtain a solution. As shown in Tables 6 and 7, while ensuring that the parameter estimation value does not diverge, this method also achieves a certain accuracy; in the second stage, as the number of discarded singular values decreases, the parameter estimation accuracy further improves; finally, in the third stage, accurate results are obtained using unbiased estimation.

Table 6 Comparison of health parameter identification results between single-stage and multistage methods

Parameter name	True value (%)	Single-stage method based on unbiased estimation (%)	Multistage method (%)		
			Stage one $m = 9$	Stage two $m = 1$	Stage three $m = 0$
ΔSW_1	– 2.04	– 0.6156	– 1.7682	– 2.0548	– 2.0392
ΔSE_1	– 1.5	19.287	– 1.3818	– 1.427	– 1.5066
ΔSW_2	– 2.08	– 24.802	– 0.3218	– 2.1999	– 2.0722
ΔSE_2	– 1.46	– 34.585	– 1.6533	– 1.5658	– 1.4535
ΔSW_{25}	– 3.91	– 3.4498	– 0.6868	– 3.2461	– 3.9494
ΔSE_{25}	– 2.94	– 2.3732	– 1.3691	– 2.925	– 2.9395
ΔSW_4	1.76	1.7231	2.2855	1.7432	1.7609
ΔSE_4	– 2.63	– 4.4766	– 2.1315	– 3.1362	– 2.5999
ΔSW_{45}	0.25	3.4073	– 1.1018	0.6137	0.2266
ΔSE_{45}	– 0.54	0.7591	– 1.4413	– 0.3385	– 0.553
Variance	–	0.2	$1.9e^{-3}$	$9e^{-5}$	$3.3e^{-7}$

Table 7 Comparison of single-stage and multistage method sensor bias error identification results

Parameter name	True value	Single-stage method based on unbiased estimation	Multistage method		
			Stage one $m = 9$	Stage two $m = 1$	Stage three $m = 0$
ΔP_{t1} (kPa)	0.1	– 6.4312	0.5354	0.0632	0.1031
ΔT_{t1} (K)	1	3.7831	0.3000	0.8994	1.0141
ΔW_f (kg/s)	– 0.0045	– 0.0368	– 0.0064	– 0.0041	– 0.0046
ΔN_1 (RPM)	– 2	– 7.2252	– 9.6869	– 2.0847	– 1.9559
ΔN_2 (RPM)	– 3	– 18.68	20.499	10.544	– 3.725
ΔP_{s25} (kPa)	0.1	– 11.796	0.496	– 0.0165	0.1097
ΔP_{t25} (kPa)	0.1	– 12.096	0.4403	– 0.0109	0.1093
ΔT_{t25} (K)	2	– 0.0733	– 0.4734	1.8044	2.0223
ΔP_{s3} (kPa)	0.1	– 58.062	2.7416	0.7249	0.0637
ΔT_{t3} (K)	– 2	– 1.7915	0.0476	– 1.5298	– 2.0094
ΔT_{t45} (K)	– 5	17.451	– 6.1546	– 5.3456	– 4.9684
ΔP_{t5} (kPa)	0.2	– 3.9187	– 0.1045	0.2398	0.1977
ΔT_{t5} (K)	5	23.05	3.22	4.8582	5.0155
Variance	–	$4.8e^3$	$6.3e^2$	$1.8e^2$	0.5

An analysis of the identification results of stage two revealed that the estimated values of the health parameters are essentially in line with the true values, but the estimated values of various sensor bias errors, such as those of the high-pressure speed N_2, total pressure after fan P_{t25}, static pressure P_{s25}, and static pressure after HPC P_{s3}, still deviate from the true values. We also observed high estimation accuracy for some parameters and low estimation accuracy for other parameters in the identification results of the single-point method. In subsequent research, we can further analyze the confidence of each parameter estimation value, which is highly valuable for engineering applications.

(3) Method of adjacent operating points—Combinations 2–4

As shown in Tables 4 and 5, the identification results for the combination of adjacent operating points 2 and 3 converge to the true value; that is, until the low-pressure conversion speed distance is 3.95%, the nonlinear identification algorithm that is based on adjacent operating points is still effective. However, when the low-pressure conversion speed distance is 3.16%, that is, in combination 4, the final identification result diverges. Tables 4 and 5 present the identification results of stage two, in which one singular value is discarded.

According to Table 3, for the same transient model operating point combinations 1–4, as the low-pressure conversion speed distance decreases, the condition number of the model-related part in the matrix $\overleftarrow{F}$ also decreases. $\underset{m}{\overleftarrow{F}}$ is defined as follows: We let $\overleftarrow{F}$ consist of $M_\theta = M_h + M_u + M_y$ column vectors:

$$\overleftarrow{F} = \left[\varphi_h^1, \ldots, \varphi_h^{M_h}, \varphi_u^1, \ldots, \varphi_u^{M_u}, \varphi_x^1, \varphi_x^2, \varphi_z^1, \ldots, \varphi_z^{M_y-2}\right] \tag{40}$$

The part related to the model is as follows:

$$\underset{m}{\overleftarrow{F}} = \left[\varphi_h^1, \ldots, \varphi_h^{M_h}, \varphi_u^1, \ldots, \varphi_u^{M_u}, \varphi_x^1, \varphi_x^2\right] \tag{41}$$

The calculation process for $\underset{m}{\overleftarrow{F}}$ uses the engine mechanism model (12), so its data collinearity level represents the similarity of the engine model at different operating points. Although the number of elements in the transient model output $\mathbf{z}$ in Eq. (12) is M_y, the high- and low-pressure speed derivatives are considered to have no measurement bias, so the number of model outputs relative to the measurement bias derivative vectors also decreases by 2, and these outputs constitute the remaining part of the matrix $\overleftarrow{F}\,\underset{z}{\overleftarrow{F}}$:

$$\underset{z}{\overleftarrow{F}} = \left[\varphi_z^1, \ldots, \varphi_z^{M_y-2}\right] \tag{42}$$

$\overleftarrow{F}_z$ is a matrix that is composed entirely of 0s and 1s, which can be directly determined according to expression (13) for the transient model output measurement value $\mathbf{z}_m$, and the process does not involve the mechanism of the engine model itself.

According to Tables 3–5, when a unified model structure (such as a transient model) is used for health parameter estimation and the condition number of the matrix $\overleftarrow{F}_m$ is greater than a certain value, the parameter estimation process does not converge because of the amplification effect on model error $\mathbf{v}$. Therefore, the condition number of $\overleftarrow{F}_m$ can be used to analyze the identifiability of health parameters. In addition, the test plan can be designed according to the model simulation results; that is, the adjacent operating points need to be selected such that the condition number of $\overleftarrow{F}_m$ is greater than a certain value.

Conclusions

This paper proposes a multistage nonlinear least squares identification method that uses the measurement data of adjacent steady-state operating points to address the issue of parameter estimation problems being underdetermined because the number of sensors available for system output is less than the number of parameters to be identified when using measurement data to estimate engine component health parameters. On the basis of the publicly available JT9D engine simulation model, after introducing health parameters and sensor measurement bias errors, the effectiveness of the proposed method is verified using simulation data.

(1) On the basis of measurement data for 3 adjacent steady-state operating points, the parameter estimation problem becomes overdetermined, and the 23 parameters to be identified, which include 10 component health parameters and 13 sensor measurement bias errors, are estimated accurately using 10 system outputs, which greatly improves the parameter identification accuracy over that of the traditional single-point method.

(2) To address the issue of severe homogenization of the engine model for adjacent operating conditions, which makes parameter identification difficult, this study uses the data collinearity analysis method from system identification theory to analyze the inherent mechanism of large deviations in parameter identification results. On this basis, a multistage nonlinear least squares identification algorithm that is based on a mixture of biased/unbiased estimates is proposed, which can effectively solve this problem. Ultimately, a high estimation accuracy can be achieved even when the low-pressure conversion speed differs by 4%.

(3) Traditional health parameter estimation methods use a steady-state engine model, whereas the method employed in this study uses a transient model. When the partial derivatives of the model are calculated with respect to health parameters and model input measurement deviations, the richness of model information is increased, thereby ultimately improving the accuracy and convergence of parameter identification.

On the basis of the above research, further in-depth research can be carried out to address the following:

(1) Although this study uses a transient engine model for parameter estimation, it uses only measurement data from steady-state operating points. The use of measurement data from transient processes can further enrich the information contained in the measurement data and the model, which would be beneficial for improving the accuracy of parameter identification. However, there is a serious problem: the impact of measurement noise. For steady-state operating points, the impact of noise can be greatly eliminated by taking averages over a period of time, but this does not hold for transient processes. At each moment, the measurement noise will have a great impact on the calculation results of the model, especially when the data collinearity is severe, which will amplify the impact of noise on the parameter estimation value. Therefore, exploring appropriate methods for addressing the adverse effects of noise is necessary.
(2) The identification results of the single-point method and the biased estimation method proposed in this paper indicate that some parameters have high estimation accuracy, whereas other parameters have low estimation accuracy. The confidence intervals of different parameter identification results are different. Therefore, in practical engineering applications, further analyzing and calculating the confidence intervals or uncertainty ranges of parameter identification results on the basis of system identification theory and methods are very meaningful.

References

1. YG Li 2002 Performance-analysis-based gas turbine diagnostics: a review Proc Inst Mech Eng Part A J Power Energy 216 5 363 377
2. Volponi AJ (2014) Gas turbine engine health management: past, present, and future trends. J Eng Gas Turbines Power 136(5)
3. Urban LA, Volponi AJ (1992) Mathematical methods of relative engine performance diagnostics. SAE Trans, 2025–2050
4. Mavris DN, Denney R (2009) Optimal robust matching of engine models to test data. Georgia Institute of Technology
5. Simon DL, Garg S (2010) Optimal tuner selection for Kalman filter-based aircraft engine performance estimation. J Eng Gas Turbines Power 132(3)
6. Chen Q, Sheng H, Zhang T (2022) An improved nonlinear onboard adaptive model for aero-engine performance control. Chin J Aeronaut

7. D Simon DL Simon 2010 Constrained Kalman filtering via density function truncation for turbofan engine health estimation Int J Syst Sci 41 2 159 171
8. A Stamatis K Mathioudakis G Berios 1991 Jet engine fault detection with discrete operating points gas path analysis J Propul Power 7 6 1043 1048
9. AG Stamatis 2011 Evaluation of gas path analysis methods for gas turbine diagnosis J Mech Sci Technol 25 469 477
10. K Mathioudakis P Kamboukos A Stamatis 2002 Turbofan performance deterioration tracking using nonlinear models and optimization techniques J Turbomach 124 4 580 587
11. Morelli EA, Klein V (2016) Aircraft system identification: theory and practice. Sunflyte Enterprises, Williamsburg, VA
12. Doel DL (1994) TEMPER—a gas-path analysis tool for commercial jet engines
13. Belsley DA, Kuh E, Welsch RE (2005) Regression diagnostics: identifying influential data and sources of collinearity. Wiley
14. Escher PC (1995) Pythia: an object-orientated gas path analysis computer program for general applications
15. Montgomery DC, Peck EA, Vining GG (2021) Introduction to linear regression analysis. Wiley
16. Sallee GP (1978) Performance deterioration based on existing (historical) data; JT9D jet engine diagnostics program
17. Sallee GP (1979) Performance deterioration based on in-service engine data: JT9D jet engine diagnostics program
18. Z Wei S Zhang S Jafari 2020 Gas turbine aero-engines real time on-board modelling: a review, research challenges, and exploring the future Prog Aerosp Sci 121 100693
19. Chapman JW, Lavelle TM, May RD et al (2014) Toolbox for the modeling and analysis of thermodynamic systems (T-MATS) user's guide

Auxiliary Generation Technology for Engine Model Code on the Basis of Prompt Engineering

Abstract The digital and intelligent development of aero engines has led to an increasing number of modeling and code development tasks. With the advancement of generative artificial intelligence technology, products such as ChatGPT and Copilot have achieved significant results in the fields of text generation and code writing. However, in professional modeling scenarios, owing to the high skill requirement for the prompt interaction process, the generated code often fails to meet the expected computational requirements. Therefore, auxiliary generation technology for model code on the basis of prompt engineering, which is based on software requirement specifications, has been used to design a prompt framework and process for the field of aero engine modeling by effectively guiding the generative model to generate code snippets that meet specific requirements. The application of this technology helps shorten the code development cycle and improve modeling efficiency, thereby promoting the development of engine digitization and intelligence. This chapter uses an engine fuel system as a typical case to show the specific application methods of the prompt framework and process for code generation.

Keywords Aero engine · Prompt engineering · Prompt framework · Auxiliary modeling · Fuel system

Introduction

In recent years, the rapid rise of generative artificial intelligence has profoundly changed the patterns of science and society. From natural language processing to image generation [1, 2] and from intelligent customer service to code writing [3], generative technology has shown strong universality and creativity and has become a core hotspot in current artificial intelligence research and applications. Large-scale pretraining models, with their extensive knowledge reserves and powerful reasoning abilities, not only provide powerful tool support for solving complex problems but also promote the full extension of artificial intelligence applications from simple tasks to more complex and higher value scenarios.

Q. Xu et al., *Digital Intelligent Aeroengine*,
https://doi.org/10.1007/978-981-95-0942-3_7

With the continuous breakthrough of large model technology, its application has gradually extended to various professional fields, such as medical diagnosis, financial analysis, and industrial design. This trend significantly enhances the level of intelligence in traditional industries and highlights the urgent need to build large models in professional fields. Compared with general models, professional-field large models give more attention to the depth of specific field knowledge and the high adaptability of tasks and can provide more accurate and efficient solutions for complex scenarios. Especially in technology-intensive industries such as aerospace, intelligent tools that are driven by generative models are gradually becoming a key force for accelerating research and development and innovation, thus laying an important foundation for the development of next-generation cutting-edge technologies.

In the development of large models in professional fields, there are two main technical routes: The first route is to integrate professional field knowledge during the training phase to build a dedicated model with domain adaptability. This method can accurately meet the needs of the field, but it requires many computing resources and high data processing costs, and the model update cycle is long. The second route combines a general large model with professional knowledge to achieve intelligent applications in a retrieval-enhanced manner. The core technology of the second route is prompt engineering, which has received much attention in recent years. Prompt engineering guides generative models to better adapt to specific field requirements while retaining general capabilities by designing and optimizing prompt language [4, 5]. Compared with directly training dedicated models, development paths that combine professional knowledge bases and prompt engineering are not only lower in cost and higher in efficiency but also have the advantage of flexibly adapting to different tasks. Therefore, the use of development paths that combine retrieval enhancement and prompt engineering is gradually becoming an important trend in the development of large models in professional fields by providing a new choice for developing intelligent tools with lower cost and higher efficiency, and the potential and value of these paths have been initially recognized in many fields.

In the field of aero engines, the application potential of large models is considerable and covers design, testing, maintenance, and other stages. In the design stage, large models can be used to quickly answer design questions (QA), assist in generating simulation models and control algorithm codes, reduce errors and improve research and development efficiency. In the testing stage, large models can generate test cases, analyze test results, accelerate the verification process, and enhance reliability. In fault diagnosis and maintenance, generative models can efficiently extract data features, combine expert knowledge, quickly locate problems, and propose solutions. These scenario concepts, to some extent, highlight the potential value of large models in the field of aero engines.

This chapter focuses on auxiliary technology for generating professional model code for aero engines. Although generative models have shown great potential in code generation, many challenges remain in their application in professional modeling fields, and the design of prompt language is particularly critical. When different prompt languages are used to express the same concept, the generated code may vary significantly, and inappropriate prompt language may even lead to results that deviate from expectations [6]. The difficulty of and skillfulness required for designing high-quality prompt language limit the effective application of the model. Owing to the research progress of prompt engineering, carefully designed prompt language has improved in regular code generation, but in complex professional modeling fields, the efficiency of generated code is still low, and meeting precision and complexity requirements is difficult [7]. Therefore, further optimizing prompt engineering has become an important direction for improving the practicality of generative models.

In research on the modeling and simulation of aircraft engine fuel systems, performance modeling is an important foundation for supporting control design optimization, integrated verification, and troubleshooting [8]. Currently, commercial software such as Simcenter dominates the research on fuel systems because of its high-precision simulation capabilities, rich visualization tools, and user-friendly interfaces [9–11]. However, the high licensing costs and limited customization capabilities of commercial software make meeting diverse needs difficult, especially in cross-platform applications and embedded systems [12]. In contrast, custom development or open-source software, owing to its flexibility and cost effectiveness, has gradually become another choice for small R&D teams and specific field research. However, the lack of maturity and maintenance significantly increases the workload of code development and model adjustment, thus posing considerable challenges for researchers.

With the rapid development of generative artificial intelligence (GAI), the problem of large amounts of modeling code development being required is expected to be alleviated. In code generation, generative models provide the possibility of improving model development efficiency and lowering the development threshold. In this context, this chapter proposes a generative model prompting method that is based on software requirement specifications. By designing a prompting framework and process, this method uses generative models to automatically generate code snippets and continuously optimizes the generated results through interaction and feedback, thereby significantly shortening the coding time and improving the modeling efficiency. This method not only provides a new technique for modeling in the field of aircraft engines but also provides practical evidence for the application of generative models in the engineering field.

The remainder of this chapter is organized as follows: Section "Prompting Framework That Is Based on Software Requirement Specifications" introduces a prompting framework that is based on software requirement specifications and proposes a generative model prompting method that is based on requirement specifications, Section "Generative Model Prompt Process for Modeling Assistance" elaborates on the generative model prompting process for assisting modeling and focuses on the process of hint design and model interaction optimization, Section "Case Study: Engine Fuel System Model Code Generation" demonstrates the application of this method in actual engineering through a case of engine fuel system model code generation, and Section "Outlook" looks forward to the development prospects of this technology and discusses the potential and challenges of generative models in the field of complex modeling. Finally, the end of the chapter provides related references.

Prompting Framework That Is Based on Software Requirement Specifications

Software requirement specifications are key documents for software development and detail the functions and performance requirements that a software system should achieve, as well as the constraints during the design and implementation process [13]. Introducing them into prompting engineering for code generation can provide important professional domain knowledge and a reference basis for hint design and is expected to improve the low effective code generation rate.

Among the many software requirement specifications, the "IEEE Recommended Practice for Software Requirement Specifications" (IEEE Std 830-1998) [14] is classic and widely used. This set of specifications not only provides detailed guidelines and standard formats for writing software requirement documents but also provides several common standardized templates for guiding development. This section designs a prompting framework for large model code generation on the basis of these software requirement specifications, as shown in Fig. 1.

The design of the above prompting framework draws on the software requirement outline of the software development document in these specifications, and a top-down layout is adopted, which includes a top-level hint area and multiple unit hint areas. The content of each hint should come from the software requirement document and should meet the constraints of the above framework, as shown below.

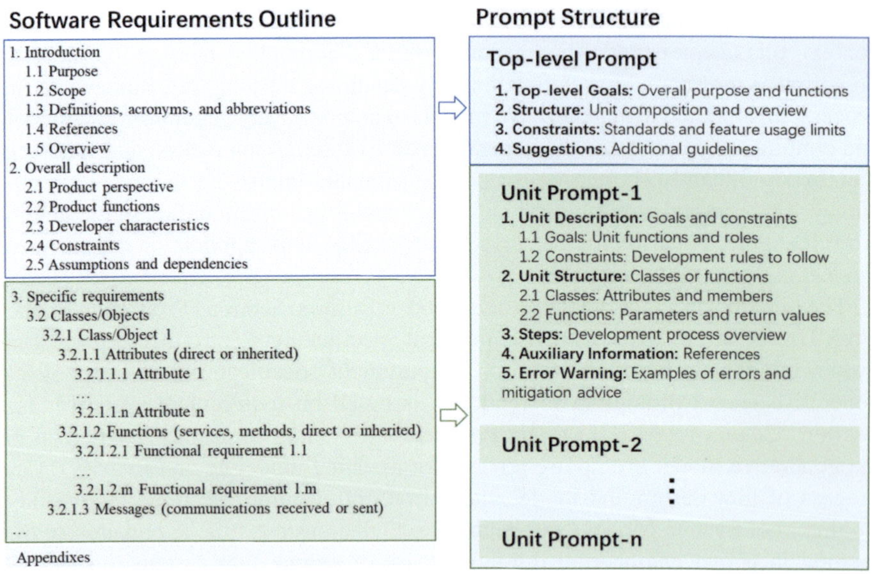

Fig. 1 Prompting framework based on software requirement specifications

Top-Level Prompts

To help the generative large model understand the content of a project's code development faster, the construction of top-level prompt language should start from an overall perspective and describe the project overview and important global information. On the basis of this idea, the designed top-level prompt area includes 4 points: "top-level goals", "organizational structure", "top-level constraints", and "other suggestions". A specific introduction to each point in the top-level prompt area is as follows.

(1) Top-Level Goals

A top-level goal is to describe and prompt at the overall purpose and function of the project by succinctly expressing the purpose of code development. For example, "This project aims to achieve the following development goals and can achieve certain functions."

(2) Organizational Structure

The organizational structure describes the development units included in the project and the role of each unit. For example, "The entire project development consists of a specified number of units, and the name and a function overview of each unit are as follows."

The types of development units include mainly classes (object-oriented) and functions (procedure-oriented), which are the basic components of the project organizational structure. In addition, the organizational structure also stipulates the generation order of each unit.

(3) Top-Level Constraints

The top-level constraint section describes the code standards to be followed during the development process and the usage restrictions on features and functions. By building prompt language—for example, "what identity to program in and what coding style to follow"—generative models can generate code in a specific coding style. On this basis, the generative model can generate code that meets the expected style and specifications, which helps reduce the extra workload caused by the use of nonstandard code, thereby improving development efficiency and code quality [15].

(4) Other suggestions

In addition to the above points, if there are other requirements or suggestions during the code generation process, they can be supplemented in the last part of the top-level prompt language according to the actual situation.

Unit Prompt

After the top-level prompt language is constructed, the unit prompt language is constructed to elaborate on the specific development requirements of each unit to assist the generative model in generating code unit by unit. The unit prompt language includes 5 points, namely, "unit description", "unit structure", "implementation steps", "auxiliary information", and "error warning", and the specific requirements are as follows:

(1) Unit Description

The unit description describes the development goals and constraints of the unit. The development goal describes the specific functions and roles of the unit, and the constraint item describes the restriction rules to be followed during the development process of the unit; for example, "the function and role of this function (development goal); the algorithm or specific variable naming rules to be adopted (constraint item)."

(2) Unit Structure

In the unit structure section, the prompt language should describe the specific structure of the class or function that implements this unit.

When the unit is a class, the properties and members of the class need to be described in detail. First, the member variables and requirements of the class are described. For example, "This class has three member variables, namely, VarA (integer), VarB (character), and VarC (real)." The functions of each member function are then explained. For example, "This class includes two member functions, Func1 and Func2, which are used for what functions."

When the unit is a function, the function's passed parameters and return values need to be described in detail, for example, "The input parameter of this function is VarA (integer); the output parameter is VarB (character); what is the meaning and type of the function return value."

(3) Implementation Steps

In the "Implementation Steps" section, the prompt language should describe the calculation logic and implementation steps inside the unit, including any important algorithms or decision-making processes; for example, "Step 1: Define variables VarA and VarB. Step 2: Use what method to calculate variable VarC. Step 3: Use what method to calculate variable VarD……".

(4) Auxiliary Information

In the "Auxiliary Information" section, prompt language can supplement the reference information related to the generation of unit code, such as parameter tables and program examples.

(5) Error Warning

The error warning section is used to provide the generative model with explanations of error handling and exceptional situations. If there is information such as compilation errors in the process of generating code, it can be added in this section or by giving examples to describe the errors that the code should avoid and the solutions. This measure helps prevent the same errors from occurring again when generating code.

Generative Model Prompt Process for Modeling Assistance

In the previous section, we introduced the prompt framework on the basis of software requirement specifications. In the process of generating code, incorporating the prompt process shown in Fig. 2 is also necessary. This prompt process includes the following 10 steps:

Step One: Obtaining the software requirement document

The software requirement document details the specific requirements for project development and is compiled according to software requirement specifications. The software requirement specifications indicate that the software requirement document is usually written by the customer and the supplier together, and these specifications also specify the source of the software requirement document.

The software requirement document contains much professional domain knowledge and is good material for professional domain code generation. Taking the example of engine model code generation, the performance requirements, functional requirements, and constraint conditions detailed in the software requirement document provide clear guidance for the generation of model code. By parsing the requirement document, key engineering parameters and control logic can be accurately extracted, thereby generating high-quality code that meets actual requirements, which is beneficial for the accuracy and stability of the model during operation.

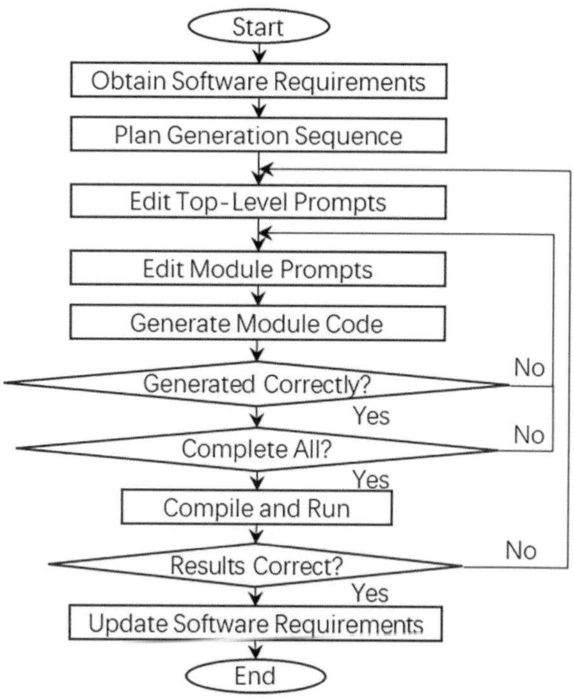

Fig. 2 Generative model prompt flowchart for code generation

Step Two: Planning the code generation order

Developers plan the order of code generation on the basis of the development content in the software requirement document to ensure the logic and coherence of the development process. Taking the example of engine model code generation, developers usually first clarify the functional requirements of the core modules, such as the thermodynamic calculation module, control logic module, and state monitoring module, and then generate code step by step according to the dependency relationships among the modules. This orderly planning not only improves the efficiency of code development but also reduces the complexity of later integration and debugging, thereby ensuring the functional consistency and performance optimization of the entire model.

Step Three: Editing the top-level prompt language

Developers edit the top-level prompt language according to the requirements of the prompt framework, that is, they complete the content involved in the 4 points mentioned in Section "Top-Level Prompts" "Top-Level Prompts": "top-level goal", "organizational structure", "top-level constraints", and "other suggestions".

Step Four: Editing the unit prompt

Developers write each unit prompt in accordance with the requirements of the prompt framework, following the planned code generation sequence. For each unit, completing the content involved in the five key points described in Section "Unit Prompt", namely, "unit description", "unit structure", "implementation steps", "auxiliary information", and "error warning", is necessary.

Step Five: Generating the unit code

Once a unit prompt is constructed, the generative model generates the corresponding code for the current unit on the basis of the top-level prompt information and the current unit's prompt information. Taking engine model code generation as an example, after clarifying the overall goal and architecture of the model in the top-level prompt information, each unit prompt is further refined to specific functional modules, such as compressor characteristic modeling, combustion chamber thermodynamic calculation, or control system logic implementation. The generative model, in combination with these prompts, gradually generates the code for each module. The constraints in the top-level prompt information help unify the code style and enhance the interface compatibility between modules.

Step Six: Determining whether the unit code has been correctly generated

Developers need to preliminarily check whether the code generated for this unit is correct. If the code is correct, they proceed to the next step; if the generated code is incorrect, the prompt of this unit continues to be optimized, and the code is regenerated. Methods for determining whether the generated code is correct include compilation verification, unit testing, and comparison with the requirement document. Compilation verification ensures the syntax correctness of the code, unit testing verifies whether the code function meets expectations, and comparison with the requirement document confirms whether the code logic and implementation meet the requirement specifications, thus ensuring the accuracy and reliability of the generated code.

Among them, compilation verification is the easiest to automate. It can quickly detect syntax errors and basic structural problems, serving as a necessary but not sufficient condition for correct coding. Although passing compilation can ensure that the code is technically executable, it cannot guarantee the correctness of functional logic or the degree of requirement satisfaction. Therefore, more comprehensive verification is needed in conjunction with unit testing and requirement comparison checks.

Step Seven: Determining whether all units have been generated

Developers should check whether all unit codes have been generated. If all units have been generated and verified, proceed to the next process; if there are still units that have not been generated, return to step four, continue to construct the prompt for the next unit and generate the corresponding code until all unit codes are successfully generated.

Step Eight: Project compilation and running

After all the unit codes are generated, the project can be compiled and run.

Step Nine: Determining whether the running results are correct

Subsequently, developers need to check whether the compilation and running results of the project are correct. For cases where the compilation fails or the running result is incorrect, return to step three, add the compilation error information to the prompt (such as the "Other Suggestions" in the top-level prompt or the "Error Warning" part of the relevant unit prompt). By adjusting the prompt to optimize it for specific error messages, the code can be regenerated to solve the problem.

Step Ten: Updating the software requirement document

The software requirement document is an important support material for code generation in the engine professional field. On the other hand, the errors and experience accumulated during the code autogeneration process can also feed back to the optimization and improvement of software requirements. By combining the problems encountered during the development process, developers can identify incomplete, unclear, or potential conflicts in the requirement description and optimize and update them. This cycle of improvement not only enhances the accuracy and operability of the requirement document but also provides clearer and more efficient guidance for subsequent development, thereby continuously optimizing the code generation process and the overall quality of the project.

Case Study: Engine Fuel System Model Code Generation

The previous sections introduced the generative model prompt method for assisting in engine model code development, which includes a prompt framework that is based on software requirement specifications and the corresponding prompt process. Notably, owing to the uncertainty of the results of the generative model, this prompt method cannot fully guarantee the correctness of the generated model results. Therefore, in actual applications, adjusting and optimizing the prompts and the generated code may be necessary to ensure that the final results meet the expected requirements.

This section uses the modeling of the aircraft engine fuel system as an example to demonstrate the specific application process of the above prompt method. This example covers the typical cases of working fluid property calculation, hydraulic component construction, and nonlinear equation solving. In the example, the large language model uses GitHub Copilot as an auxiliary tool [7].

Working Fluid Property Class

The working fluid property class is used to describe the characteristics of the fluid. It needs to be universal for different types of fluid working fluid, and it can be used to calculate physical parameters such as the density, viscosity, and specific heat of a fluid at a given temperature and pressure.

The development sequence for this case is as follows: First, the working fluid property class is defined. Then, initialization, file reading, and other class member functions are implemented. Finally, global functions for calculating fluid density, viscosity, and specific heat capacity are developed.

On this basis, the top-level prompts are constructed as shown below. The "Top-Level Goals" section clarifies the purpose of the working fluid property class; the "Organizational Structure" section provides a detailed introduction to the structure, composition, name and function of each part of the class; the "Top-level Constraints" section specifies the programming language features that cannot be used; and the "Other Suggestions" section assumes the identity of the developer and specifies the code style to be followed. By assuming the identity of the developer, the generative model can generate code according to specific coding standards, which helps reduce the workload for syntax error checking [16].

*/**

This c/c++ source file is used to read and store fluid characteristic data and to calculate fluid properties.

It has 1 class and 3 functions.

The class is called fluid_properties, which can read and store fluid characteristic data.

Three functions are used to calculate fluid properties, including density, viscosity, and specific heat.

The last function is the main function, which is used to test the fluid_properties class.

This file adheres to ISO C++ standards.

The following c++ features are not allowed in this file: namespace, operator overloading, string, and vector.

*Coding as a developer from Google and following the Google C++ style guide:*https://google.github.io/styleguide/cppguide.html

**/*

The top-level prompts provide a framework and guiding principles for the development process, which helps improve the efficiency of subsequent unit code generation. After completing the top-level prompts, unit prompts can be gradually constructed according to the planned development order.

For the definition of a class, its unit prompts should describe its attributes and member composition in detail. First, the purpose of the class and its role in the system are clarified; second, the names of the member variables in the class are listed; and finally, the type, function, and association method of each member variable with the class behavior are explained in detail. For example, the unit prompts constructed for the fuel system working fluid property class are shown below.

*/**

Describe:

This class is used to read and store one type of fluid characteristic data.

It has 4 member variables: fluid_name (char), polynomial_type (int), coefficient (double), and coefficient_name (char).

The coefficient has a fixed length of 57, which is the maximum number of coefficients for the "advanced" mode.

For the "simple" mode, only the first 24 coefficients are used.

It has 2 member functions: init and read_data.

The polynomial coefficients are listed in the following order.

**/*

Variable and function names should have clear meanings; for example, names such as "fluid_name" or "read_data" should be used, rather than generic variable names such as "VarA" from the example in the previous section. This not only improves the readability of the code but also helps the generative model better understand the context, thereby improving the accuracy and consistency of the generated code.

According to the above prompts, GitHub Copilot can automatically generate code, as shown in the Fig. 3.

In addition to defining classes, unit prompts can also be used to build functions. For the prompts of function units, the "unit description" should be provided first, which is basic information such as the function's purpose and involved methods;

```
class fluid_properties {
public:
    char fluid_name[20];
    int polynomial_type;
    double coefficient[57];
    char coefficient_name[57][20];
    int init(const char *fluid_name);
    int read_data();
};
```

Fig. 3 Code that defines the working fluid property class generated on the basis of prompts

followed by the "unit structure", which introduces the function's input and output and the parameter name, meaning, units, etc.; then the "implementation steps", which describe in detail the function's implementation steps; and finally, the "auxiliary information" and "error warnings".

For example, the unit prompts constructed for the fluid density calculation function are as follows.

*/**

Describe:

tfrhopti computes the density of a fluid by a polynomial formulation at a given pressure and temperature.

For a given index of fluid fi, the coefficients for the polynomial can be read from the TFFD3[fi] coefficient.

*The code should start with "double tfrhopti(double *p, double *temp, int *fi) {"*

Inputs:

*p, double * absolute working pressure [PaA]*

*temp, double *, working temperature [degC]*

*fi, int *, index of the thermal–hydraulic fluid*

Outputs:

void

Return value:

*double, rho, which denotes the density of the fluid [kg/m**3]*

Steps:

1. Define the parameters used in the calculation. Defining all the parameters above is not necessary; only the parameters used in the calculation are considered.

2. Calculate the difference $\Delta p = p - p0$.

3. Calculate the difference $\Delta T = T - T0$.

4. Calculate the density $\rho = \rho 0/(1 + ap2\Delta p\char`^2 + aT2\Delta T\char`^2 + apT\Delta p\Delta T + ap2T\Delta p\char`^2\Delta T + ap2T2\ \Delta p\char`^2\Delta T\char`^2 + apT2\ \Delta p\Delta T\char`^2 + ap3\ \Delta p\char`^3 + ap3T\ \Delta p\char`^3\Delta T + ap3T2\ \Delta p\char`^3\Delta T\char`^2 + aT\ \Delta T + ap\ \Delta p)$.

Common mistakes to avoid:

Input: [20]- cpT

*Incorrect output: double cpT = TFFD3[*fi].coefficient[21];*

*Correct output: double cpT = TFFD3[*fi].coefficient[20];*

Reason: The variable should be consistent with the coefficient.

**/*

The above prompts sequentially provide the "unit description", "unit structure", "implementation steps", and "error warnings" for the density calculation function. Among them, the "unit description" indicates that the function's purpose is to calculate the fluid density on the basis of the given working fluid pressure and temperature; the "unit structure" details the function parameters, return value types, and meanings; the "implementation steps" detail the specific process and required formulas for calculating the working fluid density; and the "error warnings" describe potential problems or errors that may occur during code generation. On the basis of this prompt, GitHub Copilot can automatically generate code, as shown in the Fig. 4.

```
double tfrhopti(double *p, double *temp , int *fi) {
    // Define the parametens used in the calculation
    double p0 = TFFD3[*fi].coefficient[0];
    double T0 = TFFD3[*fi].coefficient[1];
    double rho0 = TFFD3[*fi].coefficient[2];
    double ap2 = TFFD3[*fi].coefficient[14];
    double aT2 = TFFD3[*fi].coefficient[11];
    double apT = TFFD3[*fi].coefficient[13];
    double ap2T = TFFD3[*fi].coefficient[24];
    double ap2T2 = TFFD3[*fi].coefficient[25];
    double apT2 = TFFD3[*fi].coefficient[26];
    double ap3 = TFFD3[*fi].coefficient[27];
    double ap3T = TFFD3[*fi].coefficient[28];
    double ap3T2 = TFFD3[*fi].coefficient[29];
    double aT = TFFD3[*fi].coefficient[10];
    double ap = TFFD3[*fi].coefficient[12];

    // Calculate the difference Δp = p - p0
    double delta_p = *p - p0;

    // Calculate the difference ΔT = T - T0
    double delta_T = *temp - T0;
    // Calculate the density p using
    // the given polynomial formulation
    double denominator = 1.0 + ap2 * delta_p
                        * delta_p + aT2 * delta_T
                        * delta_T + apT * delta_p
                        * delta_T + ap2T * delta_p
                        * delta_p * delta_T + ap2T2
                        * delta_p * delta_p * delta_T
                        * delta_T + apT2 * delta_p
                        * delta_T * delta_T + ap3
                        * delta_p * delta_p * delta_p
                        + ap3T * delta_p * delta_p
                        * delta_p * delta_T + ap3T2
                        * delta_p * delta_p * delta_p
                        * delta_T * delta_T + aT
                        * delta_T + ap * delta_p;
    double rho = rho0 / denominator;
    return rho;
}
```

Fig. 4 Code for the density calculation function generated on the basis of the prompt

During the code generation process, if errors are encountered (such as the common error types mentioned in reference [17]), specific cases can be introduced to modify the prompt to guide the generative model to "learn from one example and apply to others" to regenerate the correct code. For example, in this case, an error similar to that described in references [18, 19] occurred during the initial generation, where the generated content did not match the requirements of the prompt. By clearly identifying the error in the prompt and switching to another generative model for generation, the problem was successfully avoided. This method falls within the research scope of "few-shot learning" [20]. It enables the model to learn and correct errors by observing a few examples. With this technique, generative models can more efficiently identify and fix problems in code generation, thus improving the quality and accuracy of generation.

Mechanical Hydraulic Components: Orifice Construction

Mechanical hydraulic components are a series of devices and components used to control, transmit, and convert hydraulic energy. Through hydraulic control, the precise metering of fuel flow or the driving of variable geometry actuators can be achieved. Hydraulic components usually include pumps, throttle valves, and electro-hydraulic conversion devices, among other components. This section takes an orifice as an example to demonstrate the automatic generation process of its unit code.

The unit prompt for generating the orifice calculation function is constructed as follows.

*/**

Describe:

orif3_ computes the flow rate through an orifice at given port pressures.

format: "orif3_(&p1, &p2, area, &hdia, &cqmax, &lamc);"

Inputs:

*p1, double *, pressure at the first port*

*p2, double *, pressure at the second port*

area, double, cross-sectional area

*hdia, double *, hydraulic diameter*

*cqmax, double *, maximum flow coefficient*

*lamc, double *, pressure at which the transition between the laminar and normal equations occurs*

Return value:

Flow rate through the orifice

Local variables:

temp, double, initial value is 25.0

fi, int, initial value is 0

Steps:

1. Define the local variable.

2. Calculate the mean pressure pmean = (p1 + p2)/2.

```
double orif3_(double *p1, double *p2, double area,
              double *hdia, double *cqmax, double *lamc)
{
    double rho, mu, dp, dpabs, lambda, cq, pmean;
    double temp = 25.0;
    int fi = 0;
    pmean = (*p1 + *p2) / 2.0;
    rho = tfrhopti(&temp, &pmean, &fi);
    mu = tfmupti(&temp, &pmean, &fi);
    dp = *p1 - *p2;
    dpabs = fabs(dp);
    lambda = *hdia / mu * sqrt(2.0 * dpabs / rho);
    cq = *cqmax * tanh(2.0 * lambda / *lamc);
    return cq * area * sqrt(2.0 * dpabs / rho);
}
```

Fig. 5 Code for the orifice calculation function generated on the basis of the prompt

3. Calculate rho and mu using the functions tfrhopti and tfmupti.
4. Define the differential pressure (absolute value).
*5. Calculate the current flow number lambda = hdia/mu * sqrt(2*dp/rho).*
*6. Calculate the flow coefficient cq = cqmax*tanh(2*lambda/lamc).*
*7. Return the flow rate cq*area*sqrt(2.0*dpabs/rho).*
**/*

The above prompts sequentially provide the "unit description", "unit structure", and "implementation steps" for the orifice calculation function. The "unit description" describes the development goal and format of the function and indicates that it calculates the volumetric flow rate through the orifice on the basis of the pressures at both ends of the orifice; the "unit structure" details the function's parameters and the types and meanings of the return values; and the "implementation steps" detail the specific steps and required formulas for calculating the volumetric flow rate through the orifice. On the basis of this prompt, GitHub Copilot can automatically generate code for the orifice calculation function, as shown in the Fig. 5.

Nonlinear Equation Solution

The calculation process for the aero engine model inevitably involves the solution of nonlinear equation groups, which requires the use of iterative solution algorithms such as the bisection method and Newton's method. This section takes the bisection method for solving a single-variable nonlinear equation as an example to demonstrate the automatic generation process of the unit code.

The unit prompt for generating the bisection method solution function is constructed as follows:

*/**

Describe:

nlsolv is used to solve a nonlinear equation, and is limited to 1 variable.

Inputs:

*xl, double *, the lower estimate of the solution*

*xr, double *, the upper estimate of the solution*

func, a function of single or double precision variables that defines the nonlinear equation to be solved

max, the maximum number of iterations to attempt

Outputs:

found, int, 0 if a solution is found; otherwise 1

*xm, double *, the solution to the nonlinear equation*

Steps:

*1. Define all local variables used in the following iterations, e.g., fl = (*func)(xl), as well as fm and fr.*

2. Initialize the variables.

3. Set the iteration tolerance tol = 1e-6 and perform the iterative computation.

**/*

On the basis of this prompt, GitHub Copilot can automatically generate the code for the throttle calculation function, as shown in the following Fig. 6.

After all unit code generation is completed, the code can be compiled and run. After appropriate debugging of the program, the running results can be obtained, as shown in Fig. 7.

Notably, the generated model code also contains test content. In the example considered in this section, the test content requires the model to output the density, specific heat capacity, and absolute viscosity of three types of fuel—Jet A1, Jet A, and Jet B—under specific conditions (a pressure of 101,325 Pa and a temperature of 25 °C).

Table 1 compares the output results of the generative model with the corresponding reference values of AMESim. The results show that the relative errors in the calculations of the density, specific heat capacity, and absolute viscosity of the three working fluid are all less than 1%. In addition, statistically, the use of traditional development methods to complete this case development requires approximately 800 keyboard inputs (close to the total number of characters in the noncomment code). In this case, the main content of the prompts comes directly from the software requirement document, and the implementation of this case development requires only approximately 40 keyboard inputs (mainly used for pasting prompts from the software requirement document and executing operations of the generative model). This method significantly shortens the development time and greatly improves the modeling efficiency.

```
void nlsolv_(double *xl, double *xm,
             double *xr, double (*func)(double *),
             int *max, int *found)
{
    double fl, fm, fr, tol;
    int i;
    fl = (*func)(xl);
    fr = (*func)(xr);
    if (fl*fr > 0.0){
        *found = 1;
        return;
    }
    tol = 1.0e-6;
    for (i = 0; i < *max; i++){
        fm = (*func)(xm);
        if (fabs(fm) < tol){
            *found = 0;
            return;
        }
        if (fl*fm < 0.0){
            *xr = *xm;
            fr = fm;
        }
        else{
            *xl = *xm;
            fl = fm;
        }
        *xm = (*xl + *xr) / 2.0;
    }
    *found = 1;
    return;
}
```

Fig. 6 The solver function code generated according to the prompt

Outlook

This chapter considered the auxiliary generation of aero engine model code and proposed a generative model prompt method based on software requirement specifications. The core content of the research included designing a prompt framework on the basis of software requirement specifications and the corresponding prompt process and verifying it through a case study of aero engine fuel system model generation. The research results show that this method can successfully guide the generative

```
powershell
Success: Open file fluid_characteristic_data/jetA1.data
Fluid name: jetA1
p0: 100000.000000
T0: 10.000000
ρ0: 806.049300
μ0: 0.001635
cp0: 1897.857518
tfrhopti: 795.568120
tfcppti: 1968.625684
tfmupti: 0.001214
Success: Open file fluid_characteristic_data/jetA.data
Fluid name: jetA
p0: 100000.000000
T0: 10.000000
ρ0: 812.035190
μ0: 0.002016
cp0: 1897.866097
tfrhopti: 801.856990
tfcppti: 1968.636792
tfmupti: 0.001456
Success: Open file fluid_characteristic_data/jetB.data
Fluid name: jetB
p0: 15000000.000000
T0: 50.000000
ρ0: 742.541881
μ0: 0.000493
cp0: 2168.947522
tfrhopti: 749.457192
tfcppti: 2080.797892
tfmupti: 0.000659
Success: Open file fluid_characteristic_data/JP-4.data
Fluid name: JP-4
p0: 15000000.000000
T0: 50.000000
ρ0: 742.541881
μ0: 0.000493
cp0: 2168.947522
tfrhopti: 749.457192
tfcppti: 2080.797892
tfmupti: 0.000659
```

Fig. 7 Running results of the engine fuel system generation model

model to generate highly targeted code, and the generated code runs correctly within the specified error range, thus meeting the modeling requirements. This provides a new idea and method for model code development in the field of aero engines.

Compared with traditional manual development methods, this method has significant advantages in terms of development efficiency and flexibility. Developers can

Table 1 Generated code test results and error analysis

Working fluid	Property category	Generated model output	AMESim reference value	Relative error %
Jet A1	Density	795.57 kg/m^3	795 kg/m^3	0.072
	Specific heat	1968.63 J/kg/K	1970 J/kg/K	0.070
	Absolute viscosity	0.0012 kg/m/s	1.21 kg/m/s	0
Jet A	Density	801.86 kg/m^3	802 kg/m^3	0.017
	Specific heat	1968.64 J/kg/K	1970 J/kg/K	0.069
	Absolute viscosity	0.0015 kg/m/s	0.0015 kg/m/s	0
Jet B	Density	749.46 kg/m^3	749 kg/m^3	0.061
	Specific heat	2080.80 J/kg/K	2080 J/kg/K	0.038
	Absolute viscosity	0.66 kg/m/s	0.66 kg/m/s	0

quickly generate code by directly selecting content from software requirement documents to construct prompts and, using generative model tools, greatly shorten the development time. However, the research process also has several shortcomings. For example, the replacement or version upgrade of the generative model may cause the same prompts to generate different code results, thus deviating from the expected model construction. This not only requires higher robustness of the prompts but also reflects that the accuracy and reliability of the generative model in complex tasks need to be further improved. Therefore, optimizing prompt strategies, enhancing the adaptability of prompt methods, and exploring ways to improve generation stability are important research directions for the future.

In the exploration of optimizing prompt strategies and enhancing prompt methods, methods based on retrieval-augmented generation (RAG) are considered an important development direction. Compared with relying solely on generative models, RAG can significantly improve the accuracy and professionalism of the generated results by combining external knowledge bases or retrieval systems. Specifically, in the scenario of auxiliary development of aero engine model code, RAG methods can dynamically call the latest data or professional experience in the domain knowledge base to ensure that the generated code meets the latest technical standards and industry norms. This combination of retrieval and generation not only effectively compensates for the shortcomings of generative models in handling complex tasks but also flexibly adapts to technological changes, thereby increasing the reliability of the generated results.

The technical solution of using a “general large model combined with RAG customization” provides a clear and feasible path for building large models in professional fields. The general large model, through large-scale pretraining, has a wide range of knowledge coverage and strong generation capabilities, whereas RAG customization is based on specific domain needs and tightly integrates the general model with specialized knowledge bases to achieve higher specificity and practicality. In the field of aero engine modeling, this solution has broad application prospects: the general large model can quickly handle general tasks, thus providing basic support for R&D, and RAG customization provides deep technical support for complex professional tasks, thereby ensuring that the generated results are more in line with industry standards.

In addition to text information, RAG can further integrate experimental data and mechanism models to provide more comprehensive support for aero engine model development. By using experimental data as input for retrieval enhancement, RAG can dynamically extract experimental results under real conditions, optimize the generated model code, and make it more in line with actual application requirements. Moreover, by applying the deductive ability of mechanism models, the flexibility of generative models can be combined with the rigor of physical laws. The generated code not only meets current needs but also has higher theoretical accuracy and engineering applicability. For example, in the modeling of aero engine fuel systems, by integrating experimental data and thermodynamic mechanism models, RAG can assist in generating more stable working fluid property calculation functions, thereby effectively reducing the debugging workload and error rate in subsequent development.

With the continuous advancement of auxiliary code generation technology in professional fields, the model construction process will gradually move from auxiliary generation to fully automatic generation. This technological evolution from "human–machine collaboration" to "full automation" can not only greatly improve development efficiency and significantly shorten the R&D cycle but also increase the creative potential of engineers by enabling them to focus on higher-level innovation and optimization work. This breakthrough development will inject strong momentum into the innovation of cutting-edge technologies in aerospace and other fields and promote the development of complex systems to reach new heights.

References

1. Sen W, Li Q, Maodi H et al (2024) A review of the research progress on question answering technology based on large language models. Data Anal Knowl Discov 8(06):16–29
2. Lee U, Jung H, Jeon Y et al (2024) Few-shot is enough: exploring ChatGPT prompt engineering method for automatic question generation in English education. Educ Inf Technol 29(9):11483–11515
3. Giray L (2023) Prompt engineering with ChatGPT: a guide for academic writers. Ann Biomed Eng 51(12):2629–2633
4. Polak MP, Morgan D (2024) Extracting accurate materials data from research papers with conversational language models and prompt engineering. Nat Commun 15(1):1569
5. Jinling S, Jianyong Z (2024) Research on the construction of query-based summary dataset AMTQFSum based on ChatGPT and prompt engineering. Data Anal Knowl Discov 8(Z1):122–132
6. Ekin S (2023) Prompt engineering for ChatGPT: a quick guide to techniques, tips, and best practices
7. Denny P, Kumar V, Giacaman N (2023) Conversing with copilot: exploring prompt engineering for solving cs1 problems using natural language. Proc 54th ACM Tech Symp Comput Sci Educ 5(1):1136–1142

8. Editorial Committee of "Aircraft Engine Design Manual". Aircraft Engine Design Manual (Volume Fifteen. Control and Fuel System) (2002). Aviation Industry Press, Beijing
9. Jian X, Shuyun W, Xiaogang W et al (2015) Simulation and verification of numerical model of aircraft engine fuel system. Gas Turbine Exp Res 28(2):41–44. https://doi.org/10.3969/j.issn.1672-2620.2015.02.008
10. Tao W, Wenwei C, Changmin L et al (2024) Parameter optimization design of constant pressure valve in aircraft engine fuel system. Aeroengine 50(02):134–138. https://doi.org/10.13477/j.cnki.aeroengine.2024.02.018.
11. Kim J, Lee J, Kim K (2019) Numerical study on the effects of fuel viscosity and density on the injection rate performance of a solenoid diesel injector based on AMESim. Fuel 256:115912
12. Shiyu Y, Yuanfang L, Xianghua X et al (2024) Steady-state characteristic simulation of aircraft engine fuel metering device. J Aero Engine Power 39(10):1–9. https://kns.cnki.net/kcms/detail/11.2297.V.20230331.1601.004.html. https://doi.org/10.13224/j.cnki.jasp.20220776
13. Ziyu L, Wanjun D, Dechun P, Pu L, Weifeng W (1998) Research on specification and implementation method of software requirement analysis based on object-oriented. Small Micro Comput Syst 07:60–64
14. Doe J (2011) Recommended practice for software requirements specifications (IEEE). IEEE, New York
15. White J, Fu Q, Hays S et al (2023) A prompt pattern catalog to enhance prompt engineering with ChatGPT. arXiv preprint arXiv:2302.11382
16. Gruda D (2024) Three ways ChatGPT helps me in my academic writing. Nature, 10
17. Wang Z, Zhou Z, Song D et al (2024) Where do large language models fail when generating code?. arXiv preprint arXiv:2406.08731
18. Zongbin S, Liya Z, Xiaoqiu L (2024) Research on material information extraction method based on local large language model and prompt engineering. Data Anal Knowl Discov 8(07):23–31
19. Li Y, Liu Y, Deng G et al (2024) Glitch tokens in large language models: categorization taxonomy and effective detection. Proc ACM Softw Eng 1(FSE):2075–2097
20. Pornprasit C, Tantithamthavorn C (2024) Fine-tuning and prompt engineering for large language models-based code review automation. Inf Softw Technol 175:107523

AI-Based Digital Maintenance Technology for Aircraft Engines

Abstract Artificial intelligence technology has become a key technology for improving the efficiency and reliability of aircraft engine maintenance in the field of engine fault diagnosis. AI-based fault diagnosis technology uses advanced algorithms such as machine learning, deep learning, and transfer learning algorithms to analyze the large amount of data generated by aircraft engines during operation to achieve early identification and accurate prediction of potential engine faults. This technology can learn fault patterns from complex data to improve the accuracy and speed of diagnosis. By real-time monitoring of an engine's operating status, this technology can not only remind maintenance personnel in a timely manner to intervene to prevent faults but also optimize maintenance plans and reduce unnecessary maintenance costs and downtime. In addition, with the continuous optimization of algorithms and the enhancement of data processing capabilities, AI-based fault diagnosis technology will become more precise and efficient, which is highly important for improving the operational safety and economy of aircraft engines.

Keywords Aircraft engine · Artificial intelligence · Fault diagnosis · Machine learning · Deep learning · Transfer learning

Introduction

Research Significance

Traditional aircraft engines usually retain large margins in design and operation to ensure flight safety. However, this approach limits the dynamic response of the engines to the external environment, and their performance changes, which not only reduces operational efficiency but also increases operating costs. Online status monitoring, fault diagnosis, and life prediction for an engine throughout its life cycle have become essential for improving flight safety and economic affordability and are also important directions for promoting the intelligent development of engines. The use of artificial intelligence technology to scientifically diagnose various fault problems

Q. Xu et al., *Digital Intelligent Aeroengine*,
https://doi.org/10.1007/978-981-95-0942-3_8

in aircraft engines not only has theoretical value but also is of urgent necessity in practice [1].

Compared with traditional methods, artificial intelligence technology, especially machine learning and deep learning, can discover complex fault patterns and make precise predictions through the analysis of sensor data and historical records. For example, models that are based on deep learning can extract fault features from vibrations, acoustic signals, etc., in real time to achieve early warning and avoid major accidents. AI-driven fault diagnosis technology can also combine predictive maintenance concepts to optimize repair plans before faults occur to reduce maintenance costs and extend engine life, thereby significantly improving the use efficiency of aviation equipment.

With the reasonable use of artificial intelligence technology, the fault diagnosis of aircraft engines is gradually changing from "passive repair" to "active maintenance". AI technology can not only accurately locate faults but also predict future possible fault trends in conjunction with prognostics and health management (PHM) technology, optimize the allocation of maintenance resources, and significantly reduce the dependence on human experience. This intelligent transformation not only provides security for the safe operation of aircraft but also provides important support for the digital and intelligent upgrade of aviation equipment, which has profound practical significance [2].

Status in China and Other Countries

Early aircraft engine maintenance technology relied mainly on visual inspection and simple parameter monitoring, such as judging the engine operating status on the basis of exhaust temperature, speed, and total vibration over-limit alarms [3]. However, the introduction of electronic hardware modules in the 1970s marked the beginning of digital maintenance technology for aircraft engines, as shown in Fig. 1. The widespread use of gas path sensors has significantly improved the information acquisition capabilities of digital maintenance systems, thus laying a solid data foundation for the development of modern digital maintenance technology [4]. At this stage, the engine health monitoring system (EHMS) gradually became a standard accessory subsystem for new aircraft engines, which transformed engine maintenance from a passive response mode to an active monitoring mode and led to breakthrough progress in terms of reliability and safety.

With improvements in computing power and the development of big data technology, the digital maintenance technology of aircraft engines was gradually improved in the 1990s with the addition of rule-based diagnostic models and expert systems, which can achieve more precise fault location and trend prediction [5, 6]. However, these systems are still limited in their data processing and analysis capabilities, and addressing complex nonlinear relationships and multivariable coupling characteristics is difficult. Since the beginning of the twenty-first century, the rapid development of artificial intelligence technology has brought new changes to the

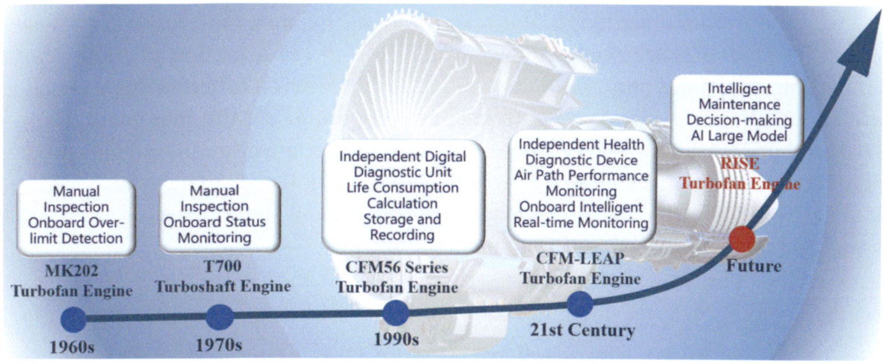

Fig. 1 Development history of aero engine operation and maintenance technology

digital maintenance of aircraft engines. In particular, machine learning algorithms, represented by deep learning, have begun to be applied in the field of engine health management. By combining historical operating data with real-time monitoring data, AI-based models can not only identify complex fault patterns but also predict future performance trends, thereby achieving precise maintenance [7, 8]. In recent years, with the development of cloud computing and edge computing technologies, the digital operation and maintenance technology of aero engines has further realized real-time data analysis and remote diagnosis, thus providing operators with more efficient maintenance solutions.

Artificial intelligence has become a core technology in the field of aero engine digital operation and maintenance; its main advantages include data-driven efficient analysis capabilities, adaptive learning capabilities, and the ability to predict complex systems. For example, through machine learning algorithms, airlines and engine manufacturers can mine early signals of engine performance degradation from massive amounts of flight data, formulate more scientific maintenance plans, and avoid flight interruptions caused by unexpected failures [9]. In addition, artificial intelligence technology is widely used in areas such as fuel efficiency optimization, component life prediction, and maintenance cost control, thus creating significant economic benefits for aviation operators [10, 11].

Mainstream aero engine manufacturers such as GE, Rolls-Royce, and Pratt & Whitney have all deployed digital operation and maintenance platforms that are based on artificial intelligence, such as GE's digital twin technology, Rolls-Royce's TotalCare service, and Pratt & Whitney's Engine Wise service. These platforms use artificial intelligence algorithms to model and simulate the operating statuses of engines to help users monitor the health of engines in real time and optimize maintenance cycles [12–14]. Moreover, in China, aero engine research institutions and enterprises are also actively developing independent artificial intelligence operation and maintenance technology, and various systems have been successfully applied to the daily maintenance of domestic engines, thereby laying a foundation for the digital transformation of China's aviation industry.

In the future, aero engine digital operation and maintenance technology based on artificial intelligence will continue to develop in the direction of intelligence, networking, and autonomy. With the continuous integration of technologies such as artificial intelligence, cloud computing, edge computing, and the Internet of Things, the operation and maintenance of aero engines will become more intelligent and efficient. In the future, this technology will not only promote the digital transformation of the aviation industry but also provide strong technical support for the sustainable development of the global aviation industry.

Aero Engine Digital Operation and Maintenance Technology Framework

Concept and Characteristics of Digital Operation and Maintenance

Aero engine digital operation and maintenance refers to the process of digitally managing and optimizing the entire lifecycle of aero engines through digital technology and intelligent means. This process is based on technologies such as the Internet of Things, cloud computing, big data analysis, and artificial intelligence (AI), in combination with digital twin models, to achieve real-time monitoring of the operating status of aero engines, fault prediction, health management, and maintenance decision support [15–17].

The core objectives of aero engine digital operation and maintenance are to analyze the operating data of aero engines intelligently; improve the operating efficiency, reliability, safety, and maintenance economy of engines; reduce operating costs; and transform maintenance work from passive repair to predictive maintenance [18] and from experience-driven transformation to data-driven transformation.

The characteristics of aero engine digital operation and maintenance include the following:

(1) **Data collection**

The technical foundation of aero engine digital operation and maintenance is data [19], and the main sources of data that airlines can obtain usually include the following:

On-wing ACARS: Through the Aircraft Communications Addressing and Reporting System (ACARS), the flight status and performance data of aero engines are transmitted in real time or with a delay to the ground data service platform (DSP) in the form of messages. The DSP further distributes the data to the aircraft operator airlines; specifically, the airlines download and store the data on a server through an account given by the OEM.

At the same time, OEM manufacturers also provide paid baseline model data services, which can tell airlines the differences between individual engines on specific

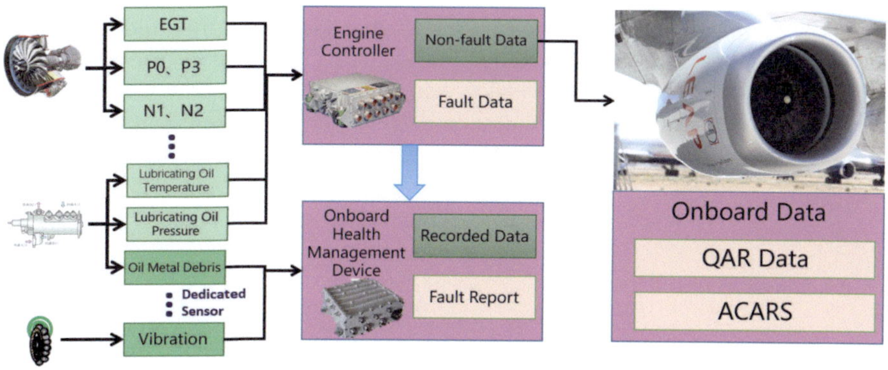

Fig. 2 Data sources for aero engine maintenance

flights and the OEM manufacturer's baseline in terms of key airway performance data. These key airway performance data generally include engine speed, exhaust temperature EGT, and fuel flow, which enable the airlines to clearly understand the degradation situation of individual engines.

QAR data: A quick access recorder (QAR) captures complete flight data, including flight parameters, engine system status, and high-frequency sampled parameter data, through multiple systems and data buses. QAR data are characterized by high accuracy and a high sampling rate and can fully reflect the dynamic changes in the engine and other systems during flight. They are usually the main data source for airlines to carry out digital operations and maintenance. Notably, airlines may regard QAR data as assets and may choose not to provide them to OEM manufacturers.

Other OEM information: Original equipment manufacturers (OEMs) usually provide various technical documents, including engine repair and maintenance manuals, service bulletins (SBs), and airworthiness directives (ADs), which constitute the technical basis for engine maintenance work. In addition, the design and maintenance of aero engines must follow the theory of thermodynamic performance cycles. The associated theoretical models are used to calculate and analyze the thermodynamic characteristics and performance parameters of the engine. These materials can also provide important references for the modeling, analysis, and decision-making of digital maintenance systems (Fig. 2).

AI-Based Digital Maintenance Framework

The AI-based digital maintenance technology framework provides comprehensive support for the digital maintenance of aero engines through data collection and fusion, intelligent diagnosis and prediction, and maintenance decision optimization [20]. Data collection and fusion solve the integration problem of multisource data and provide high-quality inputs for analysis; AI technology is used in intelligent diagnosis and prediction to achieve fault identification and performance trend prediction, thus enhancing the foresight of maintenance work; maintenance decision optimization improves maintenance efficiency and reduces costs through intelligent means, thus ultimately achieving a win–win situation in terms of economic benefits and safety (Fig. 3).

Data Collection and Fusion

Data collection is the basic process of aero engine digital maintenance. Through various sensors, data buses, and communication systems (such as ACARS, QAR, and FDR), the operating data of aero engines are obtained in real time or regularly. The types of data collected include flight parameters (such as altitude, speed, and heading), engine performance parameters (such as temperature, pressure, vibration, speed, and fuel flow), maintenance and repair records (such as historical maintenance logs, parts replacement records, and fault reports), and external environmental data (such as meteorological conditions and flight area characteristics). Together, these multidimensional data provide complete input information for the digital analysis of aero engines.

Owing to the diverse sources and inconsistent formats of the data, fusing, cleaning, and standardizing the collected data are necessary, which are often overlooked by many technicians, thereby resulting in unstable accuracy of subsequent digital maintenance algorithms. Data fusion technology can include time sequence alignment, outlier processing, and feature extraction. For example, time sequence alignment technology is used to synchronize data with different sampling frequencies to ensure data consistency; outlier processing methods are used to eliminate noise and error data during collection; and feature extraction technology is used to refine key parameters (such as health indicators and performance degradation trends) to improve data analysis efficiency and effectiveness [21].

In addition, for data collection and fusion, real-time data need to be integrated with historical data. Real-time collected data are combined with historical operating data to form a complete operating status database, which provides comprehensive support for subsequent intelligent diagnosis and prediction.

Intelligent Diagnosis and Prediction

Intelligent diagnosis is an important process in digital maintenance, which aims to analyze engine operating data through artificial intelligence (AI) technology to identify potential fault modes and abnormal trends. Fault diagnosis relies mainly on classification models, anomaly detection algorithms, and deep learning models to discover hidden fault signals in data [22–25]. For example, a classification model trained on historical data can automatically identify abnormal operating states of an

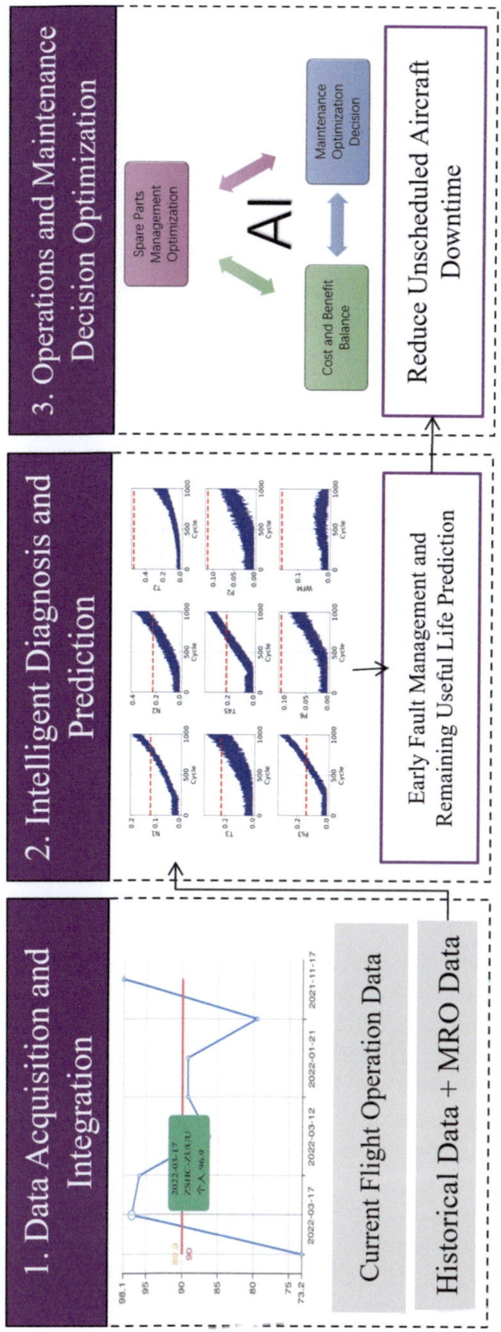

Fig. 3 Aero engine digital maintenance architecture

engine and further locate the root causes of the faults in conjunction with a fault tree or rule reasoning.

On this basis, predictive maintenance predicts the performance degradation trend and remaining useful life (RUL) of key engine components through time series analysis, deep learning, and statistical modeling methods. For example, a time series model that is based on deep learning (such as LSTM or transformer) can capture trends and changes in engine operation, thereby providing accurate degradation predictions. In addition, the combination of data-driven models and physical mechanism models further improves the accuracy and reliability of predictions.

Digital twin technology also plays an important role in intelligent diagnosis and prediction. By constructing a virtual digital model of an aero engine, its operating status can be simulated in real time, the fault development process can be predicted, and the diagnosis results can be verified. Digital twin technology makes diagnosis and prediction more intuitive and accurate, thus providing a high-value reference for maintenance decisions.

Maintenance Decision Optimization

On the basis of intelligent diagnosis and prediction, maintenance decision optimization improves maintenance plans, resource allocation, and operational efficiency through data-driven methods [26]. First, an intelligent maintenance plan optimizes the traditional regular maintenance mode via state-based on-demand maintenance. Through predictive analysis, maintenance windows can be scheduled in advance to avoid unnecessary maintenance operations while reducing unplanned downtime (aircraft on the ground (AOG)) and improving aircraft operational efficiency.

Spare part management optimization is an important part of operation and maintenance decisions. By predicting the demand for parts through AI, the spare parts inventory can be dynamically adjusted to ensure that key parts are available at any time while avoiding inventory backlog. The operation and maintenance platform can also cooperate with the supply chain system to optimize the production and transportation of spare parts, thereby improving supply efficiency.

The balance of cost and benefit is the core goal of optimizing operation and maintenance decisions. Through a comprehensive analysis of maintenance costs, flight efficiency, and engine life, a maintenance plan that maximizes economic benefits is formulated. For example, accurate fault prediction can reduce misjudged maintenance and excessive maintenance, thereby reducing maintenance costs, and optimized maintenance plans can reduce unplanned downtime and increase aircraft utilization.

Finally, operation and maintenance optimization decisions can be completed through autonomous decision-making or human–machine collaboration. At present, large language models such as ChatGPT, DeepSeek and other artificial intelligence technologies are gradually entering engineering applications. An artificial intelligence system generates an optimal decision plan on the basis of data analysis and optimization algorithms, and operation and maintenance experts can verify and adjust the results generated by AI, thus realizing a combination of expert experience and intelligent algorithms. This human–machine collaboration method not only improves the accuracy of decision-making but also enhances the flexibility and reliability of the system.

Operation and Maintenance System Technical Framework Design

The technical architecture design is divided into three key parts: a data layer, a model layer, and an application layer. Each layer supports the other layers in achieving the goals of digital operation and maintenance of aero engines. The data layer is responsible for the collection and storage of multisource data, and it provides high-quality basic data for the entire architecture; the model layer provides high-precision analysis and prediction capabilities through the combination of physical models and AI models; and the application layer considers actual operation and maintenance scenarios and supports intelligent diagnosis, fault prediction, and operation optimization by providing comprehensive technical support for airlines for improving operational efficiency, reducing maintenance costs, and ensuring flight safety.

Data Layer: Multisource Sensor Data Collection and Storage

The data layer is the foundation of the entire digital operation and maintenance technical architecture and is responsible for the collection, transmission, and storage of multisource data. Large amounts of real-time data are generated during the operation of an aero engine. These data, which include key performance parameters such as temperature, pressure, vibration, speed, and fuel flow, as well as flight status data (such as altitude, speed, and heading) and environmental data (such as meteorological conditions), are collected by sensors distributed in the engine and other systems of the aircraft.

These data are transmitted to the ground system through onboard communication systems (such as ACARS), quick access recorders (QARs), and flight data recorders (FDRs). The ground system stores and manages the data and uses distributed databases and cloud storage technologies to support large-scale, high-frequency data processing needs. To ensure data quality, the data layer also integrates functions such as data cleaning, denoising, time series alignment, and feature extraction to provide high-quality input data for subsequent analysis.

In addition, the data layer also supports the integration of real-time data streams and historical data by building a unified data storage architecture to provide comprehensive data support for upper-layer models and applications.

Model Layer: Integration of Physical Models and AI Models

The model layer is the core part of the technical architecture. It integrates the fusion capabilities of traditional physical models and artificial intelligence (AI) models and is used to analyze, simulate, and predict the operating status of aero engines. Physical models are based on the thermodynamic cycle and structural characteristics of aero engines and use a series of mathematical formulas and simulation technologies to simulate and calculate the performance of the engines. These models have a strong theoretical basis and are suitable for explaining the operating mechanism and fault mechanism of engines.

AI models, on the other hand, use data-driven methods, machine learning and deep learning technologies (long short-term memory (LSTM), transformers, random forests, etc.) to train and analyze historical data and real-time data, capture complex nonlinear relationships, identify abnormal patterns, predict performance degradation trends and remaining useful life (RUL).

The combination of physical models and AI models achieves complementary advantages. Physical models provide reliability and interpretability, whereas AI models supplement the ability to recognize complex data patterns. Through fusion, the model layer can have both theoretical guidance and data-driven capabilities, and it adapts to diverse operation and maintenance scenarios. For example, physical models can calibrate the outputs of AI models to ensure the accuracy and credibility of prediction results, and AI models can optimize the parameters of physical models through big data analysis to improve their simulation accuracy.

Application Layer: Intelligent Diagnosis, Fault Prediction, and Operation Optimization

The application layer is the user-facing part of the technical architecture and carries out functions such as intelligent diagnosis, fault prediction, and operation optimization. On the basis of the high-quality data provided by the data layer and the analytical capabilities of the model layer, the application layer realizes real-time monitoring, health assessment, fault diagnosis, performance prediction, and operation and maintenance decision support for the operation status of aero engines.

The intelligent diagnosis module uses AI algorithms to analyze the operating status of an engine in real time by identifying potential fault patterns and abnormal trends. For example, it identifies abnormal fluctuations in parameters such as temperature and vibration through classification models and anomaly detection algorithms and analyzes the root causes of faults in conjunction with knowledge graph technology. The fault prediction module, which is based on time series models or deep learning models, predicts the degradation trends of key engine components and generates health scores and remaining useful life (RUL) prediction results to support predictive maintenance.

The operation optimization module uses the results of diagnosis and prediction to further optimize the operation strategy and maintenance plan of the engine. For example, by replacing regular maintenance with predictive maintenance, unnecessary maintenance actions are reduced; by optimizing spare parts management, inventory costs are reduced; and by combining flight plans to optimize maintenance windows, unplanned downtime (AOG) is reduced. The application layer also provides a visual interface and interactive features to enable maintenance personnel to intuitively view the operating status of the engine, diagnostic results, and optimization suggestions.

Application of Artificial Intelligence Technology in Aero Engine Maintenance

Data Collection and Preprocessing

Multisource heterogeneous data, including sensor data, flight parameters, historical maintenance records, and environmental data, constitute the core data for the digital maintenance of aero engines. These data cover a wide range and have diverse formats, but they also have data quality and consistency issues. Artificial intelligence (AI) technology can achieve efficient data cleaning and feature extraction through intelligent methods when applied to multisource heterogeneous data, thus providing a reliable data foundation for subsequent diagnosis, prediction, and optimization.

The data generated during the operation of an aero engine are multisource and heterogeneous. Sensor data are real-time and accurate but often contain noise and outliers; flight parameters come from multiple systems and may have inconsistent sampling rates; and historical maintenance records are mostly unstructured data (such as text descriptions), which are difficult to use directly for model analysis. In addition, significant challenges are also faced in terms of the time alignment, format unification, and data integrity of different data sources. The processing of multisource heterogeneous data needs to balance diversity and correlation to ensure that the data accurately reflect the operating status of the engine [27].

Artificial intelligence technology can play an important role in multisource data cleaning, especially in noise filtering, outlier detection, and data completion. Traditional data cleaning methods rely on rule setting or manual processing, while AI technology can automatically learn cleaning rules from data distributions through machine learning algorithms, thus improving cleaning efficiency and accuracy. For example:

- **Outlier detection**: On the basis of anomaly detection algorithms (such as the isolation forest, support vector machine, and autoencoder algorithms), AI can identify sensor data or flight parameters that are out of the normal range and can label or correct outliers.
- **Noise filtering**: Using deep learning technology, by constructing a denoising autoencoder or filtering model, random interference noise in sensor data can be removed while retaining key feature information.

- **Data completion**: Through the interpolation and prediction functions of deep learning (such as the LSTM model), missing time series data can be completed to ensure data integrity. For example, when sensor data are lost for a certain period, AI can intelligently complete them on the basis of historical patterns and adjacent data.

After data cleaning, feature extraction is indispensable in the processing of multisource heterogeneous data. AI technology can automatically extract key features from complex raw data and capture hidden patterns and relationships in the data. For example:

- **Time series feature extraction**: For sensor and flight parameter data, AI models can extract key features, such as trends, periodic changes, and peak behaviors, from time series. By using sliding window technology and deep learning models (such as LSTM and GRUs), dynamic features related to engine operating status can be extracted.
- **Multisource data fusion feature extraction**: Through multimodal learning technology, sensor data, flight parameters, and historical maintenance records can be fused to extract related features across data sources. For example, AI can be used to discover the relationships between specific flight parameters (such as high-temperature and high-pressure working conditions) and certain fault modes in maintenance records through correlation analysis.
- **Text data feature extraction**: From the unstructured text data in historical maintenance records, AI can be used to extract key fault descriptions, maintenance measures, and frequent problems through natural language processing (NLP) technologies (such as TF-IDF and BERT), thereby converting text information into analyzable structured data.

The goal of feature extraction is to transform complex raw data into low-dimensional and highly representative information to provide high-quality input for model training and operation and maintenance decisions.

The application of artificial intelligence in multisource heterogeneous data processing not only significantly improves the efficiency of data cleaning and feature extraction but also improves the accuracy and consistency of the processing results. Compared with traditional rule-based methods, AI technology can adapt to complex and changing data environments and successfully capture hidden nonlinear patterns. In the digital operation and maintenance of aircraft engines, AI-driven data cleaning and feature extraction reduce noise interference and optimize data quality.

Fault Diagnosis Design

Aircraft engine fault diagnosis technology is key for ensuring flight safety and improving operation and maintenance efficiency. Compared with traditional methods, the introduction of artificial intelligence (AI) technology has greatly improved the level of intelligent fault diagnosis. Through deep learning and machine learning technologies, in combination with the domain knowledge of expert systems, aircraft engine fault diagnosis can achieve full-process automation and high-precision prediction from fault detection to classification.

Aircraft Engine Fault Detection Method Based on Deep Learning

Deep learning technology has significant advantages in addressing complex data patterns and large-scale time series data and has become an important method for aircraft engine fault detection. Deep learning models such as convolutional neural networks (CNNs) and recurrent neural networks (RNNs) can automatically extract features from sensor data, flight parameters, and time series and can detect potential fault signals and abnormal patterns.

CNN perform well in handling structured data, especially frequency domain data and fault signals in the form of images. For the sensor data of aircraft engines (vibration signals, acoustic signals, etc.), spectrograms can be generated through time–frequency conversion, and then CNNs can extract spatial features from the spectrograms for the identification of complex fault patterns such as bearing faults and blade cracks. For example, on the basis of signals processed by short-time Fourier transform (STFT) or wavelet transform, CNNs can automatically detect abnormal frequency components, thus providing highly accurate fault detection results.

RNNs and their improved models (such as LSTM and GRUs) can capture the dynamic change rules in time series data and are especially suitable for trend analysis of sensor data during the operation of aircraft engines. For example, an RNN can analyze continuous changes in key parameters such as engine temperature, pressure and vibration; capture abnormal trends; and achieve early fault warning. The memory mechanism of LSTM is particularly suitable for long time series data and can accurately identify gradually evolving fault signals, such as turbine degradation or combustion stability issues.

The advantage of deep learning is that it does not rely on manually extracted features but learns complex patterns automatically through large-scale training data and is suitable for real-time monitoring and big data analysis. However, deep learning models usually require large amounts of labeled data for training and have higher requirements for computing resources. In addition, the "black box" nature of the models may lead to a lack of interpretability of the diagnostic results, so combination with other methods (such as physical models) to verify the diagnostic results is necessary.

Aircraft Engine Fault Classification Model Based on Expert Systems and Machine Learning

In the field of fault classification, the combination of expert systems and machine learning methods provides an efficient and flexible solution. Expert systems rely on domain knowledge and can quickly classify known fault patterns, whereas machine learning uses a data-driven approach and has the ability to handle complex and diverse data. The combination of the two shows significant advantages in aircraft engine fault classification.

Expert systems are based on knowledge bases and rule reasoning mechanisms and can quickly classify known fault patterns through threshold judgments, logical reasoning, and other methods. For example, expert systems can directly judge the locations and types of faults on the basis of parameter thresholds set on the basis of historical experience (such as vibration exceeding a certain limit or abnormal fluctuations in fuel flow). This method is suitable for quick classification of common faults, but its adaptability to unknown fault patterns is weak, and the maintenance and update of rules depend on the experience of domain experts.

Machine learning algorithms (such as the support vector machine (SVM), random forest, and k nearest neighbor (KNN) algorithms) can train classification models on multidimensional data and automatically classify fault types. For example, by training a random forest model with sensor data, feature distributions for different fault types can be learned to achieve high-precision classification. In addition, machine learning can also group unknown fault patterns preliminarily through clustering algorithms (such as K-means) to provide a reference for further analysis.

Combining expert systems with machine learning can result in complementary advantages. For example, in the data preprocessing stage, expert systems can filter and reduce the dimensions of data on the basis of domain knowledge, thus reducing data noise and improving the training efficiency of machine learning models; machine learning models can then pass the results to expert knowledge systems for verification after classification, thereby increasing the credibility of the diagnostic results. This combination method can quickly process known faults and discover potential unknown fault patterns through data-driven methods.

The introduction of artificial intelligence technology has transformed the diagnosis of aircraft engine faults from traditional experience-driven analysis to data-driven and intelligent analysis. Deep learning performs well in complex data pattern recognition and early fault detection, whereas the combination of expert systems and machine learning provides flexible and efficient classification capabilities. By integrating these technologies, the accuracy and efficiency of aircraft engine fault diagnosis have been significantly improved while reducing maintenance costs and unplanned downtime (AOG). In the future, with the further development of artificial intelligence technology, fault diagnosis systems will become more intelligent and precise, thus providing stronger support for aviation safety and operation optimization.

Performance Prediction and Health Management

Performance prediction and health management for aircraft engines are important technical means for improving engine operating efficiency, optimizing maintenance plans, and ensuring flight safety. By introducing artificial intelligence (AI) technology, in combination with big data analysis and deep learning algorithms, we can more accurately predict the trends of key performance parameters and assess the remaining useful life (RUL) of key components. These technologies provide a solid technical foundation for the predictive maintenance and health management of aircraft engines.

Key Performance Parameter Prediction on the Basis of Time Series Analysis

The performance of an aircraft engine is determined by multiple key parameters, such as the turbine inlet temperature, fuel efficiency, and compressor pressure ratio. Changes in these parameters not only reflect the operating status of the engine but are also closely related to its health status and performance degradation. AI technology based on time series analysis can accurately predict the trends of key performance parameters through historical data and real-time data analysis, as shown in Fig. 4, thus providing support for early anomaly detection and performance optimization.

Time series prediction models are the core technologies for performance prediction. Deep learning models such as long short-term memory (LSTM) can capture the long-term trends of performance parameters, especially in scenarios of slow degradation, such as a decrease in fuel efficiency or a continuous increase in turbine temperature. The transformer model, with its efficient parallel processing capability, can quickly model long time series and is suitable for performance parameter prediction

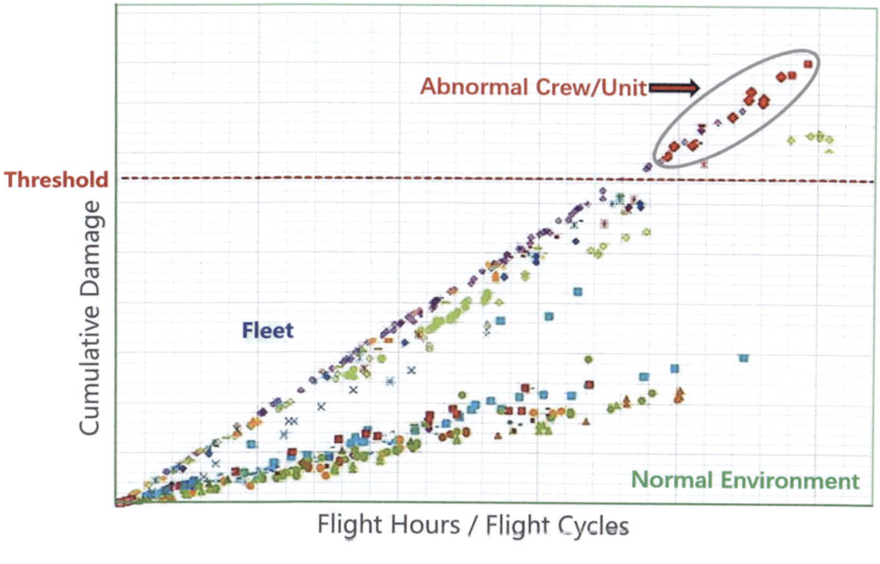

Fig. 4 Performance trend data for multiple engines of the same model

under complex conditions. In addition, multivariate time series analysis models can process multiple sensor data simultaneously to capture the dynamic relationships between different parameters, thereby improving the accuracy of prediction. For example, the change in the turbine temperature may be affected by both compressor pressure and fuel flow, and multivariate analysis can more comprehensively reflect performance changes.

Performance prediction technology that is based on time series analysis has important value in multiple operation and maintenance scenarios. For example, by predicting the trends of performance parameters, early anomaly detection can be achieved to help operation and maintenance personnel discover potential problems in advance; in flight scheduling and power distribution, optimizing fuel efficiency can reduce operating costs; in addition, these prediction results can also provide data support for maintenance plans, reduce unplanned downtime (AOG), and improve aviation operation efficiency.

Remaining Useful Life (RUL) Prediction Technology

Remaining useful life (RUL) prediction is one of the core tasks of aircraft engine health management. By assessing the degradation status of key engine components, RUL prediction can infer the remaining life values of the components in future operation, thus providing a basis for formulating predictive maintenance plans. AI technology has shown significant advantages in RUL prediction, especially in complex data modeling and dynamic change capture.

RUL prediction methods can be divided into two main categories: data-driven models and physical models. Data-driven models use historical operating data and degradation patterns to perform life prediction through machine learning or deep learning algorithms. For example, regression algorithms such as random forests and support vector regression (SVR) can estimate the life values of key components by training on sensor data and historical fault records. Deep learning models (such as LSTM and CNN-LSTM) can capture complex nonlinear degradation rules and accurately predict the life values of key components such as turbine blades and bearings. In particular, the CNN-LSTM model, which combines spatial feature extraction and time series analysis, is suitable for life assessment under complex conditions.

To increase the reliability of RUL prediction, the fusion of physical models and data-driven models is widely used. Physical models are based on the design parameters and operating mechanisms of the aero engine and have good interpretability and reliability, whereas data-driven models can dynamically adapt to the changes in actual operating conditions by learning the patterns in historical data. For example, deep learning algorithms combined with thermodynamic models can utilize both engine design parameters and degradation data to generate more accurate life prediction results. The advantage of the fusion method is that it can ensure the credibility of the prediction results and adapt to diverse needs under different working conditions.

The application value of RUL prediction is reflected in three main aspects. First, through accurate life prediction, maintenance personnel can arrange maintenance plans reasonably according to the remaining life of the components, thus avoiding premature replacement or delayed repair. Second, RUL prediction helps optimize the allocation of maintenance resources, reduce unnecessary maintenance actions, and lower maintenance costs. Finally, through real-time monitoring and life evaluation, potential risks can be detected in time to ensure flight safety.

The application of artificial intelligence technology in aero engine performance prediction and health management has significantly improved the accuracy and efficiency of prediction. The prediction of key performance parameters on the basis of time series analysis provides data support for engine performance optimization and early anomaly detection, and RUL prediction technology supports predictive maintenance and repair decisions by evaluating the life status of key components. The combination of these technologies not only reduces aviation operating costs but also reduces unplanned downtime, thereby improving the reliability and safety of engines.

Maintenance Decision Optimization

The optimization of aero engine maintenance decisions is an important means for improving operational efficiency, reducing maintenance costs, and reducing unplanned downtime (AOG). By introducing artificial intelligence (AI) technology, especially reinforcement learning and intelligent optimization algorithms, dynamic optimization of maintenance strategies and efficient allocation of maintenance resources can be achieved. The application of these technologies has transformed aero engine maintenance from experience-driven decision-making to data-driven and intelligent decision-making, thus providing more efficient and smarter solutions for aviation operations.

Maintenance Strategy Optimization Based on Reinforcement Learning

Reinforcement learning (RL) is an artificial intelligence method that learns an optimal strategy through trial and error. It shows significant advantages in optimizing aero engine maintenance strategies. Traditional maintenance strategies usually rely on maintenance at preset time intervals (such as regular maintenance) or predictive maintenance based on simple rules. Reinforcement learning can be used to optimize maintenance decisions through dynamic learning and environmental feedback, thus making the maintenance strategy more flexible and efficient.

The core idea of reinforcement learning in maintenance strategy optimization is to model the operating state of an aero engine as a Markov decision process (MDP), and through the interaction of an intelligent agent with the environment, the optimal maintenance strategy is learned. For example, we define the following:

- **State**: The operating state of the engine, which includes key performance parameters (such as the turbine exhaust temperature and vibration value), health scores, and fault risk assessments.
- **Actions**: The maintenance decision actions include continuing operation, scheduling inspection, replacing parts, and stopping maintenance.
- **Rewards**: Rewards are given on the basis of the effects of maintenance decisions, such as reducing unplanned downtime, lowering maintenance costs, and extending engine life.

Through reinforcement learning algorithms (such as Q-learning, deep Q networks (DQNs), and proximal policy optimization (PPO)), the model can explore different maintenance strategies under various working conditions and environmental conditions and find a strategy that is optimal in terms of long-term benefits. Specifically, reinforcement learning can help airlines achieve the following goals:

- **Dynamic Maintenance Optimization**: Maintenance strategies are adjusted dynamically according to the real-time health status of the engine and the operation plan to avoid over maintenance or delayed repairs.
- **Balancing Cost and Risk**: Considering the repair cost, fault risk, and operational benefits, a maintenance plan that balances short-term costs with long-term reliability should be formulated.
- **Improvement of Maintenance Efficiency**: The bias caused by human decision-making is reduced through intelligent decision support, and the efficiency and accuracy of decision-making are improved.

Intelligent Scheduling and Resource Allocation Algorithm

The maintenance process for aircraft engines involves complex task scheduling and resource allocation and includes maintenance personnel scheduling, maintenance station allocation, and spare part inventory management. Traditional scheduling and allocation methods often rely on manual allocation on the basis of experience or simple rule algorithms and have difficulty coping with complex maintenance requirements with multiple tasks and constraints. Artificial intelligence technology, through intelligent scheduling and resource allocation algorithms, can optimize operational efficiency and resource utilization while satisfying constraint conditions.

Intelligent scheduling algorithms use optimization models and AI technology to find the best order of maintenance tasks under multiple constraints. For example, heuristic algorithms (such as genetic algorithms and particle swarm optimization) can simulate natural selection or group behavior, optimize task scheduling order, and avoid task conflicts and resource idleness. For example, when multiple engines need maintenance at the same time, the genetic algorithm can find the optimal order of maintenance tasks, thus reducing the occupation time of the station. In addition, scheduling methods that are based on deep reinforcement learning can be used to simulate complex maintenance scenarios and learn the optimal task allocation strategy. For example, the deep Q network (DQN) can adjust the task execution plan dynamically according to the urgency of each task and the resources of the station to maximize task completion efficiency.

Resource allocation optimization is another key process in aircraft engine operation and maintenance. It includes the allocation of maintenance personnel, spare parts, and stations. In terms of maintenance personnel scheduling, AI algorithms can intelligently allocate tasks on the basis of personnel skills, task requirements, and time constraints. For example, classification algorithms that are based on machine learning (such as support vector machine (SVM)) can assign tasks to the most suitable technicians, thereby improving maintenance efficiency. In terms of spare part management, AI can accurately predict the demand for spare parts through predictive maintenance models and inventory optimization algorithms (such as dynamic programming), optimize the levels of spare part inventory, and avoid spare part shortages and overstocking. Station allocation optimization uses linear programming or mixed integer programming methods to optimize the order of use and allocation strategy of the station, reduce the maintenance waiting time, and improve the station utilization.

In addition, multiobjective optimization algorithms (such as Pareto optimization) can find the best balance among multiple objectives (such as reducing maintenance time, reducing costs, and improving resource utilization). For example, during peak maintenance periods, these algorithms can dynamically adjust the plan according to task priority to ensure that important tasks are completed first while minimizing the effects on other tasks.

Aircraft engine operation and maintenance decision optimization technology based on artificial intelligence, through reinforcement learning and intelligent scheduling algorithms, has achieved dynamic optimization of maintenance strategies and resource allocation. Reinforcement learning can be used to formulate the most cost-effective long-term maintenance plan on the basis of the operating status of the engine and maintenance requirements, thus helping airlines balance maintenance costs and failure risks. Intelligent scheduling and resource allocation algorithms improve task execution efficiency and resource utilization in complex multitask, multiconstraint scenarios.

Application of Digital Twin Technology in Aircraft Engine Maintenance

Digital twin technology for aircraft engines is a core technology that builds a virtual digital model of an engine, maps its physical operating status in real time, and realizes functions such as status monitoring, performance prediction, and maintenance optimization. The construction of a digital twin requires a multilevel approach in terms of structure, performance, and systems that combines physical models, data driven AI models, and multisystem collaborative simulations to fully realize the digital mapping and dynamic interaction of aircraft engines.

The construction of the digital twin of an aircraft engine covers three levels: a structural twin, a performance twin, and a system twin. The structural twin accurately describes the physical structure and mechanical behavior of the engine through 3D modeling and finite element analysis; the performance twin dynamically reflects the operating status and performance degradation trend of the engine through a combination of physical models and AI models; and the system twin fully simulates the system interaction behavior of the engine through multisystem collaborative simulation, thus providing intuitive visualization information for engine maintenance guidance.

Structural Twin: Based on 3D Modeling and Finite Element Analysis

The structural twin is the basis of the digital twin of an aircraft engine, and the core objective is to build a virtual structural model that is consistent with the real engine through 3D modeling and finite element analysis (FEA).

First, a geometric model of the aircraft engine is constructed through high-precision 3D modeling technology (such as computer-aided design (CAD), CT scanning, and laser scanning). The model needs to include key parts of the engine (such as compressors, combustion chambers, and turbine blades) and key connection structures to ensure that the geometric shape is consistent with the actual structure. In addition, material properties (such as the elastic modulus, density, and thermal expansion coefficient) and manufacturing process parameters also need to be reflected in the model to ensure the physical consistency of the model.

Second, on the basis of finite element analysis technology, multiphysics field simulations such as force, thermal, and vibration simulations are performed on the structural model. For example, during the operation of the engine, the turbine blades will bear multiple loads of high temperature and high-speed rotation. Finite element analysis can simulate the thermal stress distribution, fatigue life, and possible crack extension path of the compressor and turbine blades. These analysis results provide the structural twin with accurate mechanical behavior prediction capabilities.

The main applications of the structural twin include the following:

Structural Health Monitoring: This involves combining real-time sensor data; monitoring the stress, clearance, deformation, etc., of key components; and timely detection of abnormal situations.

Damage Assessment: Through simulation analysis of the stress concentration area or fatigue damage accumulation of the engine blades, it provides a basis for maintenance decisions. Domestic airlines such as China Eastern Airlines and China Southern Airlines have partially applied it.

Design Optimization: During the design phase, structural twinning is used for simulation verification, optimization of component design and material selection, and improving engine reliability and performance.

Performance Twinning: Combining a Physical Model with an AI Model

Performance twinning is used to construct a dynamic model of core performance parameters (turbine temperature, fuel efficiency, thrust, etc.) during engine operation that reflects the operating status and performance degradation trend of the engine. In performance twinning, the combination of physical models and AI models is key; they complement each other, ensure the theoretical consistency of the model, and enhance the prediction ability of complex nonlinear behavior.

Physical models are based on the thermodynamic and dynamic principles of aero engines, and performance outputs can be calculated by inputting operating conditions (such as speed, ambient temperature, and pressure). For example, thermodynamic cycle models can predict changes in parameters such as turbine temperature, compressor efficiency, and fuel flow. These models have high interpretability but may be limited in addressing complex nonlinear phenomena.

To compensate for the shortcomings of physical models, AI models can capture the nonlinear pattern of engine performance degradation by training on a large amount of historical operating data and sensor data. For example, time series models that are based on LSTM or transformers can predict long-term trends in fuel efficiency, and neural network regression models can accurately estimate the temperature distributions of turbine blades. In performance twinning, the integration of physical models and AI models can achieve complementary advantages. AI models can correct the parameter deviations of the physical model, and physical models provide physical constraints for the AI prediction results, thus ensuring the reasonableness of the prediction.

The applications of performance twinning include real-time monitoring of changes in key performance parameters, prediction of performance degradation trends, and optimization of engine performance (such as fuel flow and thrust distribution) in combination with operating conditions to improve operating efficiency and reduce emissions.

System Twinning: Multisystem Cosimulation

System twinning is an advanced form of digital twinning that aims to integrate multiple subsystems of an engine (such as the fuel system, cooling system, and control system) into a single model through multisystem cosimulation to fully reflect the dynamic behavior of the engine and the interactions between systems.

First, each subsystem needs to be modeled separately to reflect its characteristics. For example, the fuel system model can be used to simulate fuel flow and spray distribution, the cooling system model can be used to analyze the flow of cooling air and heat transfer, and the control system model can be used to simulate functions such as thrust adjustment and fuel injection control on the basis of the logic of the engine electronic control unit (ECU).

After completing the subsystem modeling, the system twin implements multi-system cosimulation through data interfaces. The coupling relationships between subsystems are dynamically analyzed through simulation. For example, the injection status of the fuel system affects the temperature distribution in the combustion chamber, which affects the heat exchange demand of the cooling system. Through cosimulation, complex interactive behaviors among systems can be analyzed, and the overall performance under different working conditions can be evaluated.

In addition, system twinning can achieve real-time simulation and feedback. By receiving sensor data, system twinning can simulate the dynamic behavior of an engine under current operating conditions and provide optimization suggestions. For example, in an emergency, system twinning can quickly simulate various response plans to help make the best decision. The main applications of system twinning include optimizing parameter matching between systems (such as cooptimization of fuel injection and cooling air volume), quickly locating the sources of faults, evaluating the scopes of fault effects, and analyzing the operating status of the engine under extreme conditions.

Future Development Trends of Digital Operation and Maintenance

With the rapid development of the aviation industry and the continuous progress of artificial intelligence technology, aviation engine maintenance is moving in more intelligent, digital, and collaborative directions. Future development trends will focus on the innovative application of artificial intelligence technology, the in-depth development of digital twin technology, and the comprehensive construction of an aviation engine maintenance ecosystem. These trends will drive the transformation of aviation engine maintenance from single technology breakthroughs to systematic innovations, thus providing a new impetus for high-quality industry development.

New Trends of Artificial Intelligence Technology in Aviation Engine Maintenance

With the continuous evolution of artificial intelligence technology, its application in aviation engine maintenance is following new development directions.

First, federated learning and privacy protection technology will become an important trend. In the aviation maintenance scenario, data are distributed across multiple independent nodes (such as airlines, maintenance organizations, manufacturers, etc.), and data privacy and security are often key obstacles to cooperation. Federated learning achieves multiparty joint modeling without sharing original data, which can fully utilize dispersed data resources while protecting data privacy. The application of this technology will greatly increase the breadth and depth of data analysis in aviation engine maintenance and promote cross-enterprise, cross-organized collaborative innovation.

Second, self-supervised learning and small sample learning will play an important role in processing aviation engine data. Owing to the complexity and safety requirements of aviation engines, fault data are often scarce, and the limitations of traditional deep learning, which relies on a large amount of labeled data, are particularly prominent in the aviation field. Self-supervised learning can effectively improve the learning ability of the model by mining potential patterns in massive amounts of unlabeled data; small sample learning uses a small amount of labeled data to achieve efficient modeling, which is particularly suitable for the prediction and diagnosis of rare faults. The application of these technologies will further overcome the current bottlenecks of AI technology in aviation maintenance scenarios.

Deep Development of Digital Twin Technology

The future development of digital twin technology will gradually shift from static modeling to dynamic interaction. Digital twin technology has begun to play a greater role in cross-system, multiscenario collaborative simulation, especially in recent years, as major airlines have begun to build their own digital maintenance centers for engineering applications.

First, future digital twins will transition from static twins to dynamic twins. The current digital twins are based mainly on the static model of the design stage, and future development will give more attention to real-time interactivity. Through dynamic integration with sensor data and operational data, a digital twin can reflect the operating status of an aviation engine in real time and initiate dynamic responses to performance changes and fault risks. This technological breakthrough in dynamic twinning will significantly enhance the state awareness and prediction capabilities of aviation engines, thus providing strong support for intelligent maintenance.

Second, digital twin technology will develop toward a collaborative twin model across systems and multiple scenarios. Aviation engine maintenance involves not only individual components or systems but also the coupling relationships between the entire engine and other systems (such as flight control systems and environmental monitoring systems). In the future, digital twin technology will be used to build cross-system, multiscenario twin models through multisystem collaborative simulation to fully reflect system behavior under complex operating conditions. This collaborative twinning can not only optimize the maintenance of a single engine but also provide decision support for the operation optimization of the entire aircraft.

Construction of an Aviation Engine Maintenance Ecosystem

The future development of aviation engine maintenance requires a comprehensive maintenance ecosystem that is based on a data platform to fully implement intelligent maintenance through industry collaborative innovation.

First, the development of a data platform and standardization is the foundation for building a maintenance ecosystem. In aviation engine maintenance, data sources are diverse. The data include sensor data, maintenance records, and flight logs, and the data formats and interfaces between different organizations and equipment often lack uniformity. In the future, a unified data platform needs to be established to promote the standardized development of data formats, interface protocols, and analysis standards. Through standardized data sharing mechanisms, cross-organization and cross-domain data interconnection can be achieved to provide basic support for industry collaboration and technological innovation.

Second, intelligent maintenance and industry collaborative innovation will become the core driving forces of the maintenance ecosystem. The maintenance of aviation engines involves multiple entities, such as manufacturers, airlines, and maintenance organizations. In the future, deep integration of the entire industry chain needs to be achieved through intelligent technology and collaborative innovation. For example, through an intelligent maintenance platform that is based on artificial intelligence, the operating status of an engine can be monitored in real time to provide performance improvement suggestions for manufacturers, optimize maintenance plans for airlines, and provide precise fault troubleshooting schemes for maintenance organizations. This multiparty collaborative model will significantly improve the efficiency and reliability of aviation maintenance and promote the overall development of the industry toward intelligence.

In the future, artificial intelligence technology, digital twin technology, and the construction of maintenance ecosystems will become the three core development directions in the field of aviation engine maintenance. Emerging technologies such as federated learning and small sample learning will solve data privacy and scarcity issues, thereby promoting the deep application of AI technology in aviation maintenance; dynamic twinning and cross-system collaborative twinning will further enhance the state awareness and complex system optimization capabilities of aviation engines; and the construction of data platforms and industry collaborative innovation will provide a solid foundation for intelligent maintenance to achieve maximum value across the entire industry chain. These development trends will push aviation engine maintenance toward a new stage of intelligence, digitization, and collaboration, which will inject strong momentum into the future development of the aviation industry.

Conclusions

AI-based aviation engine digital operation and maintenance technologies have significant value and meaning. These technologies improve the accuracy of engine operation status monitoring through intelligent fault diagnosis and prediction methods, thus providing the possibility to discover potential problems in advance and avoid major safety risks. In addition, performance prediction and health management technologies have transformed maintenance work from passive response to active intervention, thus significantly reducing unplanned downtime (AOG) and increasing the economic benefits of aviation operations.

The application of digital twin technology further realizes the deep integration of virtual reality, thereby enabling the operation statuses of aviation engines to be mapped and dynamically analyzed in real time in the digital space. This not only provides strong support for fault prediction, performance optimization, and emergency response but also provides a new direction for future intelligent design and manufacturing. Moreover, intelligent scheduling and resource allocation algorithms improve operational and maintenance efficiency and resource utilization by optimizing the allocation of maintenance resources, thus helping comprehensively promote the digital transformation of aviation operations and maintenance.

Overall, AI-driven aviation engine digital operation and maintenance technologies organically combine safety, reliability, and economy to provide technical support for the high-quality development of the aviation industry.

References

1. Jing L, Boyao Z, Dayi Z, Min C (2022) Current status and prospects of fault diagnosis research on aero gas turbine engines. J Aeronaut 43(08):7–20
2. Donghui L, Jiaqi Y, Mei Y (2022) Research on fault diagnosis of aero engines. Equip Manag Maint 05:148–149
3. Volponi AJ (2013) Gas turbine engine health management: past, present and future trends. American Society of Mechanical Engineers Digital Collection
4. Li YG (2002) Performance-analysis-based gas turbine diagnostics: a review. Proc Inst Mech Eng Part A J Power Energy 216(5):363–377
5. Xunkai Y (2014) Prediction and health management of aero engines. National Defense Industry Press, Beijing
6. Ming C, Jinqun H, Jian Z, Xuefeng C, Feng L, Fang W (2021) Current status, challenges and opportunities of fault diagnosis and health management of civil aero engines—fault diagnosis and prediction of gas path, mechanical and FADEC systems. J Aeronaut, 1–35

7. Tahan M, Tsoutsanis E, Muhammad M, Abdul Karim ZA (2017) Performance-based health monitoring, diagnostics and prognostics for condition-based maintenance of gas turbines: a review. Appl Energy 198:122–144
8. Shaochen L, Min C, Jintao H, Hailong T (2022) Progress in gas path fault diagnosis of aero gas turbine engines. Aero Engine 48(02):33–49
9. Tolani D, Yasar M, Chin S, Ray A (2005) Anomaly detection for health management of aircraft gas turbine engines: 2005 American Control Conference, ACC. Proc Am Control Conf 1(WeA15.1):459–464
10. Zuomin F (2004) Introduction to fault diagnosis of aero engines. Science Press
11. Jinqun H, Qihang W, Feng L (2020) Current status and prospects of gas path fault diagnosis research for aero engines. J Nanjing Univ Aeronaut Astronaut 52(04):507–522
12. Frank PM (1996) Analytical and qualitative model-based fault diagnosis—a survey and some new results. Eur J Control 2(1):6–28
13. Improved engine health monitoring using full flight data and companion engine information (2018). In: Diagnostics and prognostics of aerospace engines. SAE, pp 143–165
14. Zhou X, Lu F, Huang J (2019) Fault diagnosis based on measurement reconstruction of HPT exit pressure for turbofan engine. Chin J Aeronaut 32(5):1156–1170
15. Chang X, Huang J, Lu F (2017) Health parameter estimation with second-order sliding mode observer for a turbofan engine. Energies 10(7):1040
16. Chang X, Huang J, Lu F (2017) Robust in-flight sensor fault diagnostics for aircraft engine based on sliding mode observers. Sensors 17(4):835
17. Lu F, Zheng W, Huang J, Feng M (2016) Life cycle performance estimation and in-flight health monitoring for gas turbine engine. J Dyn Syst Meas Control 138(9)
18. Urban LA (1973) Gas path analysis applied to turbine engine condition monitoring. J Aircr 10(7):400–406
19. Urban LA (1969) Gas turbine engine parameter interrelationships, 2nd edn. Hamilton Standard Division of United Aircraft Corporation
20. Doel DL (1994) TEMPER—a gas-path analysis tool for commercial jet engines. J Eng Gas Turbines Power 116(1):82–89
21. Brotherton T, Volponi A, Luppold R, Simon DL (2003) eSTORM: enhanced self tuning on-board real-time engine model. In: 2003 IEEE aerospace conference proceedings (Cat. No.03TH8652)
22. Seo D-H, Roh T-S, Choi D-W (2009) Defect diagnostics of gas turbine engine using hybrid SVM-ANN with module system in off-design condition. J Mech Sci Technol 23(3):677–685
23. Hongwei W, Guo C, Libo C, Ke S, Ai L (2014) A kind of aviation engine rolling bearing wear fault monitoring technology. J Aerosp Power 29(09):2256–2263
24. Yanshan S, Hao Y, Haibin T, Wen Z, Zhoumo Z (2017) Online monitoring of aviation engine oil particle wear. J Instrum Meter 38(07):1561–1569
25. Weili W, Dali C (2021) Aviation engine vibration fault diagnosis technology and development trend. Aerosp Power 03:69–72
26. Jia Q, Zhao Q (2006) A SVM-based method for engine maintenance strategy optimization
27. Woike M, Abdul-Aziz A, Oza N, Matthews B (2013) New sensors and techniques for the structural health monitoring of propulsion systems. Sci World J 2013:596506

Data-Driven Intelligence Empowers Next-Generation Aircraft Engine Innovation

Abstract With the rapid development of artificial intelligence (AI), traditional big data technology is experiencing a new intelligent upgrade in the application of aircraft engine design and manufacturing. The core algorithms of AI rely on high-quality, multidimensional big data as the foundation, and big data enable deeper insights and decision-making capabilities to be obtained through AI. This chapter discusses how to drive engine design, research and development, and manufacturing efficiency and innovation breakthroughs through big data-driven AI technology and how to promote the efficient operation of aircraft engine design optimization, intelligent manufacturing, and full lifecycle management. This chapter also introduces related cases to show how big data empower AI and ultimately accelerate the transformation of aircraft engines toward "digital intelligence". Finally, this chapter looks forward to the future development trends of big data and AI in the field of aircraft engines. It provides readers with a comprehensive and forward-looking perspective and helps them take initiative in the wave of digital intelligence transformation while always remembering that data constitute the cornerstone of AI innovation.

Keywords Artificial intelligence · Data driven · Big data · Digital twin · Aircraft engine

Introduction

The performance of an aircraft engine, which is the core power device of an aircraft, directly affects flight safety, fuel economy, operating costs, and the environment. Over the past few decades, the design and manufacturing processes for aircraft engines have relied mainly on accumulated experience, physical experiments, and engineering calculations. However, with the development of technology and changes in market demand, the traditional R&D model gradually faces many problems, such as high R&D costs, long testing and verification cycles, complex manufacturing processes, and limited innovation space. To address these challenges, data-driven

Q. Xu et al., *Digital Intelligent Aeroengine*,
https://doi.org/10.1007/978-981-95-0942-3_9

Fig. 1 Data collected by the aircraft engine that reflect its status. Sourced from the AI

intelligent technology is gradually entering the field of aircraft engines and has had profound effects on design, production, maintenance, and other aspects (Fig. 1).

The application of data and intelligent technology has transformed all aspects of aircraft engine R&D from experience- and experiment-driven to data- and algorithm-driven, which has led to new design ideas and manufacturing methods. Through the synergistic effects of big data, artificial intelligence (AI), the Internet of Things (IoT) [1], digital twins [2] and other technologies, innovative design, intelligent manufacturing, and precise maintenance of aircraft engines are entering a new era. Especially in the operation and health management of engines, real-time data collection and analysis can provide more accurate health predictions and fault diagnoses for aircrafts, thus reducing the downtime and maintenance costs of airlines and operators.

This chapter focuses on how data and intelligence are deeply integrated into all aspects of aircraft engines and promotes the innovation and transformation of the entire industry, especially in the application of design optimization, intelligent manufacturing, lifecycle management, and future green intelligent engine R&D.

Data-Driven Design and Optimization

In the design phase of modern aircraft engines, the combination of data analysis, artificial intelligence (AI), and big data technology provides engineers with unprecedented design insights. This innovative technical method not only accelerates the design cycle but also greatly improves the performance and efficiency of engines. Traditional design methods usually rely on experience accumulation and many trial and error processes; however, data analysis and machine learning (ML) can quickly identify problems in design and optimize design parameters, thus making engine design more precise, efficient, and innovative. In this way, engineers can more accurately predict engine behavior, optimize each design link, and effectively respond to the increasingly stringent requirements of the aviation industry for safety, energy savings, and environmental protection.

Application of Big Data in Design Optimization

Aircraft engine design is a complex process that involves multiple disciplinary fields and usually requires highly accurate calculations and verification of aerodynamic performance, thermodynamics, material properties, structural strength, and other aspects. Each design decision may have profound effects on the performance, cost, and reliability of the engine. Therefore, extracting valuable information from massive amounts of data has become the key to improving design efficiency and accuracy. The introduction of big data technology provides strong support for this goal (Fig. 2).

Traditional engine design relies on multiple trials and accumulated experience, where designers continuously adjust the geometry, materials, and structure of components to achieve the desired performance. However, this process is very tedious and costly, especially when engine components (turbine blades, compressor blades, etc.) are optimized. Traditional trial and error methods often require long periods of experimentation and simulation, which pose significant challenges in terms of time and budget. The introduction of big data has fundamentally changed all this.

Through big data technology, designers can analyze historical design data, experimental data, material performance data, etc., in detail and discover potential design problems or optimization spaces. For example, in the design of engine turbine blades, multiple factors need to be considered, such as aerodynamic performance, thermal load, and structural strength. In traditional design, engineers need to fine-tune the geometric shape of the turbine blades through repeated trials and experiments to improve aerodynamic efficiency and the ability to withstand thermal loads. However, through big data technology, the design team can analyze many design schemes

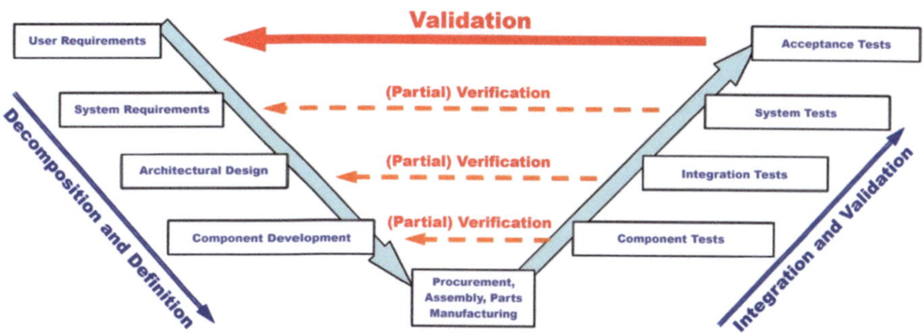

Fig. 2 Traditional V model development process [3]

through simulation technology and optimization algorithms and quickly screen out the best design schemes. This not only reduces the cost of trials but also greatly shortens the design cycle [4].

Example: A Case Study of Cooperation Between an Airline and an Engine Manufacturer

An airline cooperated with its engine supplier to collect and analyze the engine operation data of hundreds of its past aircraft, and the fatigue behavior and thermal stress distribution pattern of the turbine blades were identified. The analysis revealed that under high-temperature and high-pressure conditions, the fatigue accumulation and thermal stress concentration of turbine blades were the main causes of blade damage. With these data, the design team could better understand the load situation of the turbine blades in actual use and improve the design scheme accordingly (Fig. 3).

On this basis, the designers optimized the geometric shape of the turbine blades and chose composite materials that are more adaptable to high-temperature environments. The improved turbine blades not only achieved improvements in aerodynamic performance and fuel efficiency but also required a significantly reduced maintenance frequency due to their higher temperature resistance. This improvement greatly enhanced the reliability and fuel economy of the engine.

Advantages:

- **Improved design accuracy**: Big data enables designers to better understand all variables in the design process, thus reducing manual prediction errors.

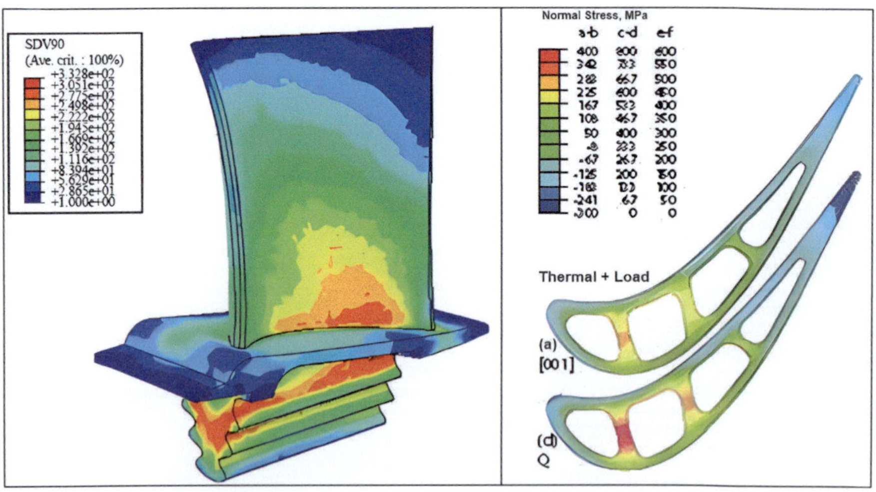

Fig. 3 Turbine blade thermal stress distribution on the basis of engine operation data. Sourced from the internet

- **Optimized design efficiency**: By using data analysis and simulation, multiple design schemes can be obtained in a shorter time, thereby greatly improving design efficiency.
- **Reduced costs**: Data-driven optimization reduces the number of tedious physical experiments in traditional design, thus saving costs.

AI-Driven Innovative Design Methods

Artificial intelligence (AI) technology, especially machine learning (ML) and deep learning (DL) algorithms, provides innovative solutions for aircraft engine design [5]. AI technology learns and discovers patterns from a large amount of historical data, thus providing a new perspective for traditional design methods. The advantage of AI lies in its powerful data analysis capability: it can identify potential patterns and rules that are difficult to detect in the traditional design process, thereby providing designers with more innovative design schemes.

Traditional engine design is usually optimized by engineers on the basis of experience, simulation results, and test data. However, owing to the complexity and high coupling of design variables, traditional design methods sometimes struggle to find the best solution quickly. AI technology can efficiently search the design space under the guidance of historical data through machine learning models, thereby discovering potential optimal design schemes.

Example: Application of AI in a Turbine Blade Design Project

In a turbine blade design project, the engineering team used AI technology to evaluate the aerodynamic performance of different blade designs. The design of turbine blades is a complex multi-objective optimization problem in which multiple indicators, such as the aerodynamic performance, material characteristics, thermal stress distribution, and mechanical strength, need to be considered. The design team built a deep learning model to learn from a large amount of wind tunnel experimental data, and combined with CFD simulation models, the AI system could quickly predict the aerodynamic performance of different blade shapes under flight conditions (Fig. 4).

Through the AI system, designers can automatically generate a variety of design schemes and evaluate the aerodynamic performance and thermodynamic efficiency of each scheme. Compared with traditional step-by-step optimization methods, AI can generate efficient design schemes in a shorter time and adjust designs according to real-time feedback. For example, the AI system can simulate the working environments of different blade shapes and material combinations; calculate the thrust output, aerodynamic efficiency, and thermal load of each design under different operating conditions in real time; and accurately predict the pros and cons of each design.

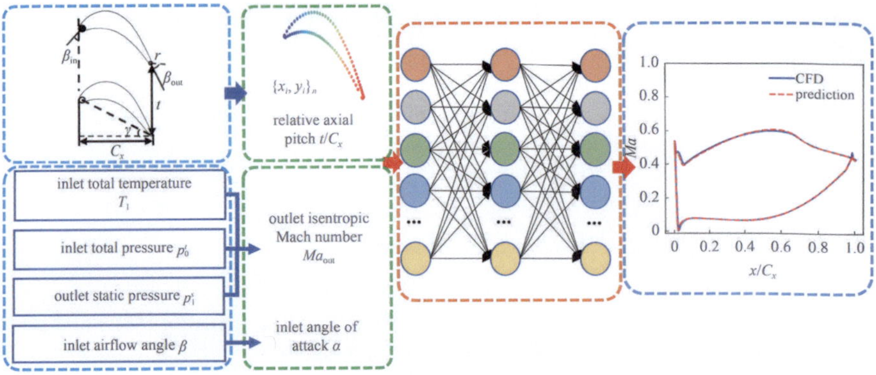

Fig. 4 Deep learning prediction of blade aerodynamic performance [6]

AI technology is not limited only to aerodynamic design but can also be applied to material selection. AI can analyze a large amount of material performance data through machine learning and select the most suitable materials for use under specific flight conditions. For example, AI-based algorithms can screen out alloy materials suitable for high-temperature environments, thus optimizing engine durability under extreme conditions.

Advantages:

- **Accelerates the design process**: AI can quickly generate and evaluate a variety of design schemes, thereby significantly shortening the design time.
- **Optimizes design decisions**: AI technology helps designers discover design patterns that are difficult to detect with traditional methods, thus enhancing the innovativeness and effectiveness of the design.
- **Reduces manufacturing costs**: By optimizing design and material selection, AI can reduce manufacturing difficulties and costs, thus ensuring that the design meets performance requirements and is manufacturable.

Case Study: Use of AI by an Aircraft Engine Manufacturer to Optimize Turbine Blade Design

As the aviation industry's requirements for engine performance, efficiency, and reliability continue to increase, traditional engine design methods are becoming no longer able to meet modern aviation needs. To overcome the design bottleneck and improve engine performance, a leading aircraft engine manufacturer used artificial intelligence (AI) technology, especially deep learning and optimization algorithms, to optimize the design of turbine blades. This innovative application not only effectively improved the design efficiency but also significantly improved the performance of the turbine blades, thereby improving the fuel efficiency and operational reliability of the entire engine [7].

Background

Turbine blades are among the most critical components of an aircraft engine, and their performance directly affects the fuel efficiency, thrust output, and safety of the engine. In the design of turbine blades, factors such as material selection, thermal stress distribution, aerodynamics, and mechanical strength usually need to be considered. These factors are often mutually restrictive. Moreover, traditional design methods require many experiments and calculations, and the trial and error process is often time-consuming and costly.

To optimize the design process, an aircraft engine manufacturer decided to use AI technology to assist in the design, especially in the optimization of the geometric shape and aerodynamic performance of the turbine blades. Through AI-optimized design, the company hopes to find an optimal solution that considers performance, materials, cost, and manufacturing processes, thereby improving the overall efficiency of the engine and shortening the development cycle (Fig. 5).

AI-Optimized Turbine Blade Design

1. **Data-driven design input**: First, large amounts of historical data and experimental data, including the geometric shapes of turbine blades designed in the past, performance parameters (pressure, temperature, airflow speed, etc.), material performance data, manufacturing process data, and performance feedback data, need to be input into the AI system. Through data cleaning and standardization, the system can quickly understand the relationships between different design variables and final performance. These data provide a basis for training the AI model.

   In addition, engineers have also obtained aerodynamic performance and thermal stress distribution data for different turbine blade designs under different working conditions through computational fluid dynamics (CFD) simulation models and structural analysis models (FEMs). These simulation data provide an important reference for AI optimization.
2. **Combination of deep learning and optimization algorithms**: In the optimization of turbine blade design, the AI system, uses deep learning and genetic algorithms and other optimization techniques to generate the optimal geometric shape. Through deep learning algorithms, the AI system can learn the relationships between various design variables and performance from a large amount of historical design data and automatically identify which geometric parameters have the greatest impact on blade performance.

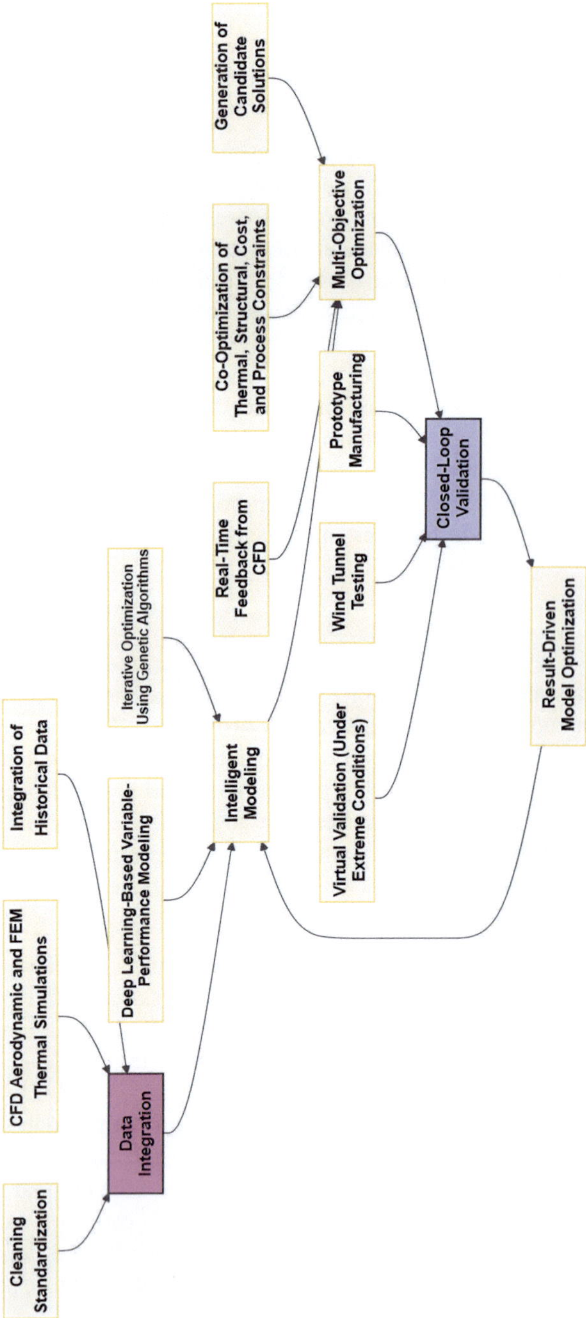

Fig. 5 Application process diagram of AI optimized turbine blade design

On this basis, genetic algorithms are used for the optimization search. Genetic algorithms simulate the process of natural selection by crossover, mutation, and selection operations and, through multiple iterations, continuously optimize the geometric shape of the blade. In each iteration process, the AI system evaluates the design results on the basis of multiple objective functions, such as aerodynamic performance, thermal efficiency, and material cost, to quickly find the best design scheme.

3. **Multi-objective optimization and simulation evaluation**: To ensure that the designed turbine blade meets the performance requirements of the aero engine and can be adapted to complex manufacturing processes, the AI optimization process uses multiobjective optimization methods. This means that the design needs to consider not only to performance but also to manufacturing processes, material use, cost control, and other factors.

   During the optimization process, the AI system, through integration with CFD simulation tools, can evaluate the airflow performance of each design variant in real time under actual flight conditions and feed the evaluation results back to the optimization algorithm. For example, the AI system can calculate the fuel efficiency and thrust output under different blade angles and airflow channel designs and further optimize them on this basis.
4. **Generation of the final design and verification**: After multiple rounds of optimization, the AI system generates multiple alternative turbine blade design schemes. Each scheme undergoes detailed simulation and analysis to ensure that its performance under extreme conditions, such as high temperature, high pressure, and complex airflow, meets the design standards of aero engines.

   After the best design scheme is selected, the engineering team makes physical models and conducts actual wind tunnel tests to verify the aerodynamic performance and structural strength of the AI-optimized design. Although this process relies on intelligent optimization by the AI system, physical experiments are still needed to confirm the performance of the design under actual conditions.

Achievements and Advantages

1. **Improved fuel efficiency**: Through AI optimization, the design of the turbine blade significantly improves the aerodynamic performance of the engine, especially in terms of combustion efficiency and thrust output. The optimized turbine blade can use air flow more effectively, reduce energy loss, and thus improve fuel efficiency. The results show that, compared with the traditional design, the turbine blade designed by AI optimization has approximately 5% higher fuel efficiency.
2. **Shortened design cycle**: Traditional turbine blade design often requires multiple rounds of design, testing, improvement, and verification, which is time-consuming and costly. AI-optimized design, through automated parameter search and optimization processes, greatly shortens the design cycle. The design team

can obtain multiple feasible design schemes within a few weeks, thus greatly improving work efficiency.

3. **Reduced material and manufacturing costs**: AI-optimized design not only focuses on aerodynamic performance but also considers the cost of materials and the feasibility of manufacturing. By optimizing the geometric shape and structure of the blade, the AI design scheme enables the turbine blade to be composed of more economical materials while meeting performance requirements and makes it easier to manufacture, thereby reducing waste and processing difficulty. Overall, manufacturing costs are reduced by approximately 8%.
4. **Improved product reliability and safety**: Through precise thermal stress analysis and structural optimization, the performance of the turbine blade designed by AI optimization under high-temperature and high-pressure conditions is improved. The new design not only improves the engine thrust and fuel efficiency but also reduces the risk of blade damage. The AI system enhances the reliability of the turbine blade through continuous simulation and verification, thereby improving the safety of the entire engine.
5. **Promotion of sustainable development**: The AI-optimized design also improves the thermal efficiency of the engine, thereby reducing fuel consumption and carbon emissions. With the aviation industry's increasing attention to environmental protection and sustainable development, the use of AI technology for optimized design helps airlines reduce their carbon footprint and achieve more environmentally friendly aviation operations.

Future Prospects

The successful application of AI in optimizing turbine blade design has given aero engine manufacturers a significant technical advantage, and this innovative application has had a profound impact on industry. In the future, the application of AI in engine design will continue, which will not be limited only to turbine blades but also extended to the optimization design of the entire engine.

With the continuous advancement of AI technology, computing power, and simulation tools, in future engine design, predicting and optimizing every design detail more accurately will be possible, thus further enhancing engine performance and reliability. Moreover, with the integration of artificial intelligence and manufacturing technology, AI will play more important roles in intelligent manufacturing, personalized customization, and real-time production optimization.

In addition, the combination of AI technology with other cutting-edge technologies, such as digital twins and the Internet of Things, will further promote the intelligent and digital transformation of aero engine design. This will make engine design more flexible and personalized to meet different customer and market needs and make a greater contribution to the sustainable development of the aviation industry.

Summary

By using AI technology to optimize turbine blade design, an aircraft engine manufacturer has not only improved the performance and fuel efficiency of the engine but also significantly shortened the design cycle and reduced costs. The successful application of AI-optimized design has promoted the innovation of engine manufacturing processes and brought more efficient, green, and reliable solutions to the aviation industry. This case study demonstrates the considerable potential of AI in the field of aircraft engines. In the future, with the continuous advancement of technology, AI will play key roles in more design and manufacturing links, thereby further promoting the intelligent and digital transformation of the aviation industry.

Intelligent Manufacturing: From Tradition to the Future

The manufacturing link is the core part of aircraft engine production and directly affects the performance, quality, and reliability of the product. With the rapid development of intelligent manufacturing technology, the traditional aircraft engine manufacturing model is undergoing profound changes. The widespread application of AI and big data not only improves production efficiency but also greatly enhances manufacturing precision and product quality. By analyzing real-time data during the manufacturing process, AI can achieve intelligent monitoring and optimization of the production process, thus making the manufacturing process of aircraft engines more refined and intelligent. The following section introduces two key areas of intelligent manufacturing: data integration and intelligent monitoring, and automation and digital twinning.

Data Integration and Intelligent Monitoring

In the traditional aircraft engine manufacturing process, production equipment and process management are often relatively independent and lack integration with and feedback from real-time data. This "island-style" management model often leads to low production efficiency and even affects the quality and reliability of the product. During the manufacturing process, the operating status of the equipment, process parameters, and production data may affect the final quality of the product; however, if this information cannot be monitored and fed back in real time during the manufacturing process, identifying problems in time and addressing them will be impossible. Therefore, data integration and real-time monitoring have become key technologies for improving manufacturing efficiency and product quality.

With the development of Internet of Things (IoT) technology, various devices, sensors, and machines in the manufacturing link can collect data in real time and

Fig. 6 Engine monitoring on the basis of the IoT and big data. Sourced from the internet

upload these data to the central management system. On the basis of the data collected by these devices, including temperature, pressure, vibration, speed, accuracy and other key process parameters, manufacturers can fully understand the real-time status of each link in the production process. Through big data analysis and AI algorithms, the production system can automatically identify abnormalities in production, such as equipment failures, process deviations, and production bottlenecks, and then perform timely adjustments and optimizations. More importantly, AI can predict equipment failures or production deviations on the basis of historical data and trends, thereby enabling preventive measures to be taken in advance to avoid failures, reduce downtime, and improve production efficiency (Fig. 6).

Example: Application of Intelligent Monitoring and Data Integration

An aircraft engine manufacturer achieved precise monitoring of the entire process of turbine blade manufacturing by adopting an intelligent production monitoring system that is based on the IoT and big data technology [8]. In this system, sensors that are installed on key production equipment can monitor the production status of turbine blades in real time and record key data, such as temperature, pressure, and vibration, in each manufacturing step. All collected data are uploaded to the central control platform in real time and analyzed by AI algorithms in real time. These data can not only reflect the current status of the production process but also reveal potential problems. For example, if the temperature of a certain production step exceeds the predetermined threshold, the system can automatically remind the operator to adjust it, and if an abnormal vibration frequency of a certain device is found, the system will issue a fault warning and carry out equipment maintenance in advance.

Through this intelligent monitoring system, problems on the production line can be identified and solved in the shortest amount of time possible, thus avoiding substandard products and the need for rework to address production deviations. Moreover, on the basis of the analysis of historical data, the system can also provide suggestions for optimizing the production process, thus helping manufacturers achieve continuous improvements in production efficiency.

Advantages:

- **Real-time monitoring**: By real-time monitoring of manufacturing process data, problems in production can be identified and solved at any time to avoid substandard products or downtime.
- **Early warning**: AI technology is used to analyze trends in changes in historical data and real-time data, predict equipment failures and production deviations, and take preventive measures in advance.
- **Improvement of efficiency and quality**: The introduction of data integration and intelligent monitoring makes the production process more precise, reduces interference from human factors, and improves the consistency and quality of the product.

Automation and Digital Twin Technology

Digital twin technology has emerged in the manufacturing industry in recent years. It involves the creation of a virtual digital model on the basis of real-time operation data of physical devices, thereby achieving comprehensive monitoring and management of the equipment. Digital twin technology can help companies not only understand the operating statuses of physical devices in real time but also simulate tests in a virtual environment, analyze the effects of different production schemes, and make optimized decisions for the production process.

In the manufacturing of aircraft engines, digital twin technology is widely used for process monitoring, optimization, and simulation [9]. For example, an aircraft engine manufacturer uses digital twin technology in the production process of turbine blades. This technology connects a digital model of a turbine blade with the real-time data of the production equipment, which enables the manufacturer to track the manufacturing status of the turbine blade in real time. Each turbine blade manufacturing process generates a virtual replica, and through the digital twin model, the manufacturer can determine in real time whether there are production deviations in the turbine blade, such as dimensional deviations and uneven material, during the manufacturing process (Fig. 7).

Fig. 7 Digital twin of an engine. Sourced from the internet

Example: Application of Digital Twin Technology

In the manufacturing process of turbine blades, digital twin technology transmits the working data (temperature, pressure, processing speed, accuracy, etc.) of the equipment in real time and feeds this information back to the virtual model in real time. Manufacturers can monitor the entire production process through the digital twin model and determine the difference between the actual production process and a predetermined target. When a deviation is detected, the virtual model can quickly alert the production staff to make adjustments. For example, if the size error of the turbine blade exceeds the allowable range at a certain stage, the digital twin model will issue an alarm, locate the deviation to a specific device on the production line, and adjust the operating parameters in time to ensure the accuracy of the final product.

In addition, digital twin technology can also optimize the production process by simulating different production schemes [10]. Manufacturers can conduct "experiments" in a virtual environment, evaluate the effects of various processes and production configurations on product quality, and choose the most suitable production scheme. This process not only reduces the number of actual physical tests and costs but also enables large-scale experiments and optimizations to be conducted without interfering with actual production.

Advantages:

- **Real-time monitoring and adjustment**: Digital twin technology achieves comprehensive monitoring and instant adjustment of the production process through a combination of virtual models and real-time data, thus ensuring product accuracy.
- **Virtual simulation and optimization**: By simulating different production schemes, digital twin technology helps manufacturers find the optimal production process, reduce costs, and improve efficiency.
- **Reduction of risk and cost**: Digital twins can be used to discover production deviations and potential risks in a virtual environment in advance, thereby avoiding major losses in actual production and reducing the cost of physical testing.

Case Study: Application of Digital Twin Technology in Engine Component Manufacturing

With the continuous development of digital technology, digital twin technology has become an indispensable key technology in aircraft engine manufacturing. Digital twin technology creates a digital model of physical objects and synchronizes their operating statuses in real time, thus achieving real-time monitoring, analysis, and optimization of actual products. This technology not only improves manufacturing accuracy and reduces costs but also accelerates innovation and product quality improvement.

Background

A globally renowned aircraft engine manufacturer that is committed to improving the manufacturing accuracy and production efficiency for turbine blades faced deformation and quality issues that were caused by complex processes such as heat treatment and casting during the production process. Traditional detection methods often rely on physical detection and sampling inspection, which not only are time-consuming and laborious but also cannot accurately evaluate each component, thereby leading to potential risks of nonconforming products and high rework costs.

To overcome this bottleneck, the company applied digital twin technology to the manufacturing process of turbine blades, created a digital model of the blade, and monitored every link in the blade production process in real time, thereby achieving precise control and quality optimization [11].

Application of Digital Twin Technology

1. **Creating a digital twin model**: In the manufacturing process, each turbine blade is first modeled using high-precision 3D scanning technology. Scanners and sensors collect geometric data on the blade in real time, convert these data into a digitized three-dimensional model, and compare these data with design data and simulation data to form the preliminary framework of the digital twin. In this way, the digital model can accurately match the actual turbine blade in terms of size, shape, etc.
2. **Real-time monitoring of the production process**: During the manufacturing process, the turbine blade needs to go through multiple stages, such as casting, machining, and heat treatment. During these processes, physical parameters such as temperature, pressure, and vibration may affect the shape of the blade. By embedding sensors into the production equipment, these parameters can be collected in real time and uploaded to the cloud platform. The data are compared with and updated on the basis of the digital twin model of the blade in real time to ensure that the geometric shape of the blade is in an ideal state at each production stage.

   For example, during the heat treatment stage, by comparing the digital twin model with the actual blade's temperature and shape change data, the system can predict in real time whether the blade will deform due to excessive temperature changes. If a deviation occurs, the system will immediately issue an alarm and adjust the processing parameters to avoid unqualified deformation of the blade.
3. **Simulation and prediction**: Digital twins are used not only for real-time monitoring but also to simulate and predict the behavior of turbine blades under different manufacturing processes in advance. For example, the system can simulate the thermal expansion effect of a blade under high-temperature conditions, combine it with the thermal stress data of the material, and predict whether cracks or other damage will occur. If risks are predicted, manufacturing personnel can take measures in advance, such as adjusting the process or optimizing the material, to reduce the costs of rework and maintenance later.
4. **Full lifecycle tracking**: After manufacturing, the digital twin model of each turbine blade continues to be used to track the status of the blade while it is in use. By integration into the engine's health management system, the digital model of the blade synchronizes the pressure, temperature, and vibration data collected during the flight process in real time. When the blade enters the later stage of use, digital twin technology can help predict its remaining service life, arrange maintenance and replacement in advance, and avoid unnecessary failures (Fig. 8).

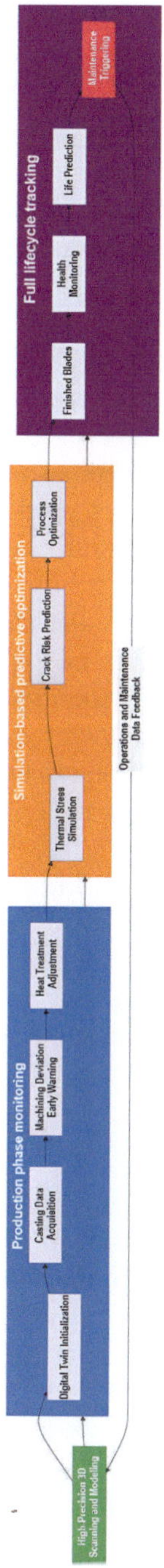

Fig. 8 Digital twin-driven turbine blade manufacturing process

Results and Advantages

By introducing digital twin technology, this aircraft engine manufacturer significantly improved the production accuracy and quality control capabilities for turbine blades, with the following specific results:

1. **Improving production accuracy**: The digital twin model can reflect the status of the turbine blade in real time, and each link in the manufacturing process can be accurately monitored, which greatly reduces the effects of human factors on product quality and ensures the high consistency of the geometric dimensions of each blade with the design.
2. **Reducing unqualified products and rework**: Through real-time monitoring and prediction, the system can detect possible problems in the production process of the turbine blade, such as deformations caused by high temperature, in a timely manner. Compared with traditional methods, digital twin technology greatly reduces the occurrence rate of unqualified products and reduces rework costs.
3. **Shortening the production cycle**: Digital twin technology can automatically identify and optimize various parameters in the manufacturing process, which reduces the time for process adjustment and the need for manual intervention, thereby shortening the production cycle and improving production efficiency.
4. **Reducing maintenance costs**: During the use of the blade, digital twin technology provides maintenance personnel with accurate health management data, predicts possible failure risks, and arranges maintenance and replacement in advance. This predictive maintenance greatly reduces the repair costs and aircraft downtime caused by sudden failures.
5. **Promoting innovation and continuous improvement**: Through the comprehensive collection and analysis of manufacturing data, digital twin technology provides the R&D team with rich data support, which enables them to quickly evaluate the effects of various designs and production processes, thereby accelerating the innovation and optimization of new technologies and products.

Future Prospects

With the continuous maturity and popularization of digital twin technology, aircraft engine manufacturers will further strengthen its application in the production of key components such as turbine blades. In the future, digital twins will not only play a role in a single manufacturing link but will also be integrated deeply with technologies such as artificial intelligence, big data analysis, and the Internet of Things to form an intelligent manufacturing ecosystem. This system can not only optimize the production stage but also achieve data synchronization and intelligent decision-making throughout the life cycle, thus promoting the development of the aviation engine industry in a more intelligent, efficient, and green direction.

In summary, the application of digital twin technology not only improves the precision and efficiency of aircraft engine manufacturing but also provides important technical support for promoting the intelligent transformation of the entire aviation industry. With the continuous progress of digital technology, we believe that future aircraft engine manufacturing will rely more on digital twin technology and intelligent means to achieve more reliable, safe and efficient aviation power systems.

Data Lifecycle Management: Data-Driven Intelligent Decision-Making

In the whole life cycle management process of aircraft engines, data play a crucial role. From design and manufacturing to testing, and maintenance, each link generates a large amount of high-value data. Efficiently accessing, integrating and utilizing these data have become essential for improving engine performance, reducing costs, and ensuring safety. With the continuous development of big data, Internet of Things (IoT) and artificial intelligence (AI) technologies, the aviation engine industry has ushered in the era of data-driven intelligent decision-making. In this process, full-type data management (including access to and processing of heterogeneous data sources) is playing an increasingly important role. This article discusses in detail how to promote intelligent decision support in the field of aircraft engines through full-type data management and thus achieve a "data-driven" transformation.

Accessing Heterogeneous Data Sources: Opportunities and Challenges

The production, use and maintenance of aircraft engines involve multiple stages, each of which generates data of different types, formats, sources and timeliness, which leads to many challenges in data management. These data sources include CAD drawings and simulation data in the design stage, process parameters and quality control data in the manufacturing stage, test data in the test stage, operation data in the use stage, and failure records and maintenance data in the maintenance stage. The data at each stage not only differ greatly in the way that they are generated and their content but also are often scattered across different systems and platforms, with each using different data formats, storage methods and update frequencies. For example, design data are usually structured CAD drawings and simulation models, whereas quality control data obtained during manufacturing are dynamic data collected in real time by sensors. The differences in the format and timeliness of these data pose great challenges when aggregating and integrating them. The heterogeneity of these data not only makes the uniform access of data from different sources difficult but also makes collaboration and data sharing between different systems and platforms more complex.

To achieve data integration and sharing, the aircraft engine industry often needs to adopt various technical means, such as standardized interfaces, middleware data platforms, and data warehouses, to solve problems such as inconsistent data formats and different system interface standards. For example, design data and manufacturing data often need to be converted and integrated through a middleware data platform so that data from different sources can be shared on the same platform. The role of the middleware data platform is to process heterogeneous data sources uniformly to provide a consistent data foundation for subsequent data analysis, processing and decision-making. Moreover, the adoption of standardized interfaces and data formats can greatly reduce the conversion costs and errors caused by inconsistent data formats, thereby improving the accuracy and usability of the data. Despite this, the implementation of these technical methods still faces considerable challenges, especially in the aircraft engine industry, where the data update frequency is high and real-time requirements are strong. Traditional data access and conversion methods often fail to meet the industry's growing needs (Table 1).

In addition, traditional practices often rely on manual data conversion and integration, which not only increases the complexity of data management but also reduces the timeliness of data analysis. Manual data conversion not only requires considerable human intervention but also easily leads to data loss or inconsistency owing to human errors. Moreover, with the dramatic increase in data volume in the production and operation process of aero engines, traditional processing methods can no longer meet the needs of real-time data processing and quick decision-making. Therefore, achieving automated integration and real-time analysis of data has become the key to promoting the digital transformation of aero engines. For this reason, an increasing number of companies are beginning to use data processing and analysis methods that are based on cloud computing, big data platforms, and artificial intelligence technology to achieve efficient and accurate data management and promote cross-departmental and cross-system data sharing and collaboration. The application of this technology helps eliminate the bottlenecks of traditional data management methods, thereby improving the data utilization efficiency and the level of intelligent decision-making on a broader level [12].

Example: Challenges of Heterogeneous Data Sources Integration

A large aeroengine company encountered problems with multiple system data that were not directly compatible when design, manufacturing, and maintenance data were integrated. The CAD drawings and simulation data used by the design department were generated by specialized engineering software, whereas the manufacturing department used an independent production management system and different databases for data storage. In addition, the maintenance department used an independent maintenance management system. When these data were integrated, problems

Table 1 Access and challenges of heterogeneous data sources in the engine manufacturing process

Stage	Data types	Data sources	Data formats	Challenges	Technical means
Design stage	CAD drawings, simulation data	Design documents, simulation software	Structured data, drawings, models	Inconsistent data formats, dispersed systems	Standardized interfaces, data conversion platform
Manufacturing stage	Process parameters, quality control data	Production line, real-time sensor collection	Dynamic data, real-time updates	Poor data timeliness, difficult to convert	Middleware data platform, standardized interface
Trial phase	Trial data	Test equipment, experimental records	Sensor data, experimental reports	Different data formats, complex data integration	Data warehouse, standardized interface
Usage phase	Operational data	Equipment operation, monitoring system	Real-time data, sensor data	Large data volume, high real-time requirements	Big data platform, cloud computing technology, automated processing
Maintenance phase	Fault records, maintenance data	Maintenance records, equipment fault detection	Fault reports, maintenance history data	High data update frequency, complex processing	Artificial intelligence technology, automated data integration

such as inconsistent data formats and nonunified interfaces occurred, which slowed the integration process.

To solve this problem, the company adopted a data middleware technology architecture. This architecture integrates data from different systems through standardized data interfaces and manages the data in a standardized manner through a middleware data platform. This not only solves the compatibility problem between different data sources but also realizes the sharing and intercommunication of data in various links, thereby providing a reliable basis for intelligent decision-making.

Challenge summary:

- Differences in data source formats and system interfaces make data integration difficult.
- Achieving interconnection and interoperability between systems through standardized interfaces and data middleware is necessary.
- The integration process of heterogeneous data sources requires considerable time and resources.

Framework and Value of Comprehensive Data Management

The core goal of comprehensive data management is to manage the data of multiple links, such as links in the design, manufacturing, operation, and maintenance of aero engines, in a unified manner to realize the comprehensive use of the data and ultimately provide support for intelligent decision-making. This framework can not only help enhance the utilization value of data but also provide accurate and timely information on the design, production, operation, and other links, thereby optimizing the decision-making process at each stage.

The challenges faced by comprehensive data management include data islands, inconsistent data quality, insufficient real-time analysis capabilities, and the growing demand for advanced analysis techniques. These challenges require us to go beyond traditional data warehouses and simple data analysis and turn to a powerful platform that can seamlessly integrate, analyze, and provide insights through artificial intelligence algorithms. Architecturally, the comprehensive data integrated management analysis system adopts a layered design, which consists of a data access layer, data analysis layer, data processing layer, metadata fusion layer, intelligent agent docking layer, scheduling decision layer, etc. It uses advanced data weaving technology to ensure that data from different sources can be efficiently integrated and associated. Moreover, built-in machine learning and artificial intelligence algorithms support automated data processing and insight generation.

Key technologies include data mapping and transformation technology, which are used to handle structural differences between heterogeneous data sources; data linking technology, which enhances data consistency and integrity by identifying and associating similar or identical entities; noninvasive multisubsystem joint processing technology, which is used to achieve highly scalable system architecture and functions; and the integration of machine learning runtime environments and database runtime environments, which reduces the cost of data flow between processing and intelligent analysis. In addition, machine learning algorithms play an important role in data weaving, especially in automatically identifying data patterns and optimizing data integration strategies (Figs. 9 and 10).

Integrated management of all types of data is the frontier direction in the field of data management. There are few systems that can achieve this goal, which include mainly UC Berkeley's Lakehouse [13] and Tsinghua's independently developed IGinX system [14, 15]. Lakehouse mainly achieves lake-warehouse integration through effective management of metadata, whereas Tsinghua's IGinX can support unified metadata management of heterogeneous data, provide noninvasive real-time access technology to achieve real-time access to heterogeneous data sources, and support efficient "data-intelligence integrated" running environments for intelligent algorithm near-number operations to greatly reduce the cost of data flow between data processing and intelligent analysis.

The main application scenarios of integrated management of all types of data are wide ranging. The integrated management of all types of data has strong potential in cross-domain knowledge discovery and complex problem solving in fields aside from

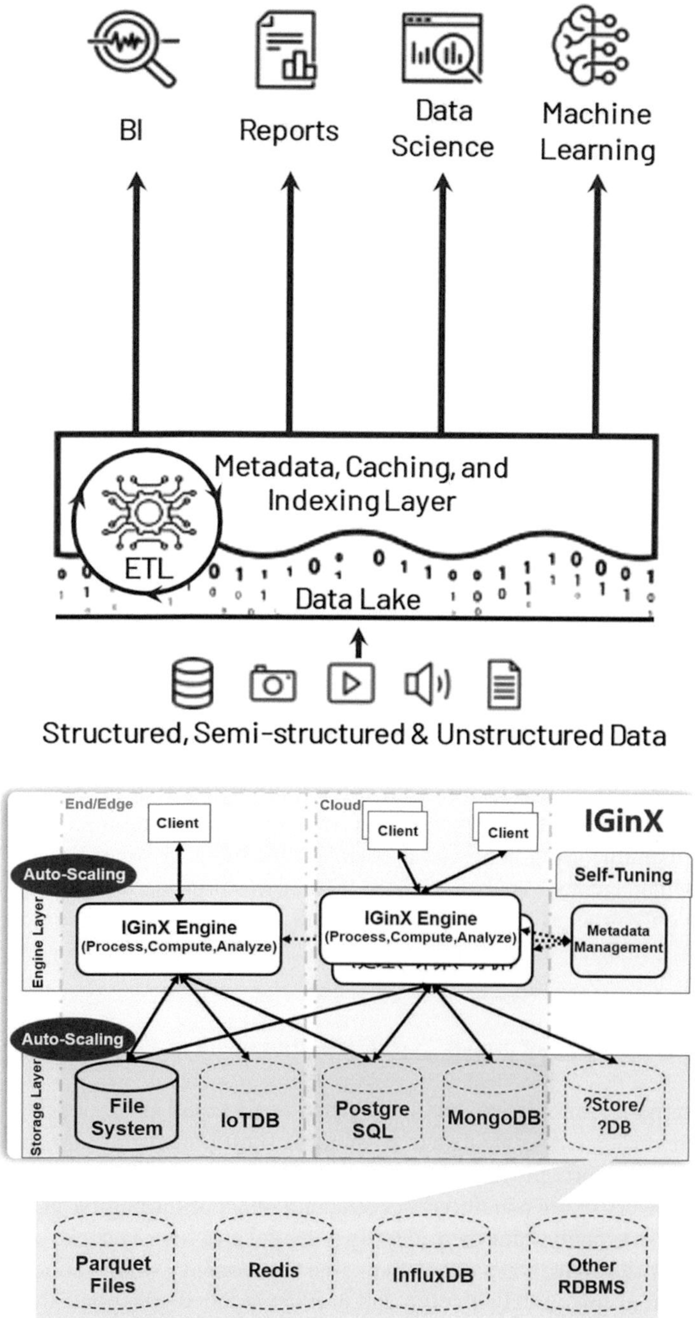

Fig. 9 Lakehouse lake warehouse integrated architecture

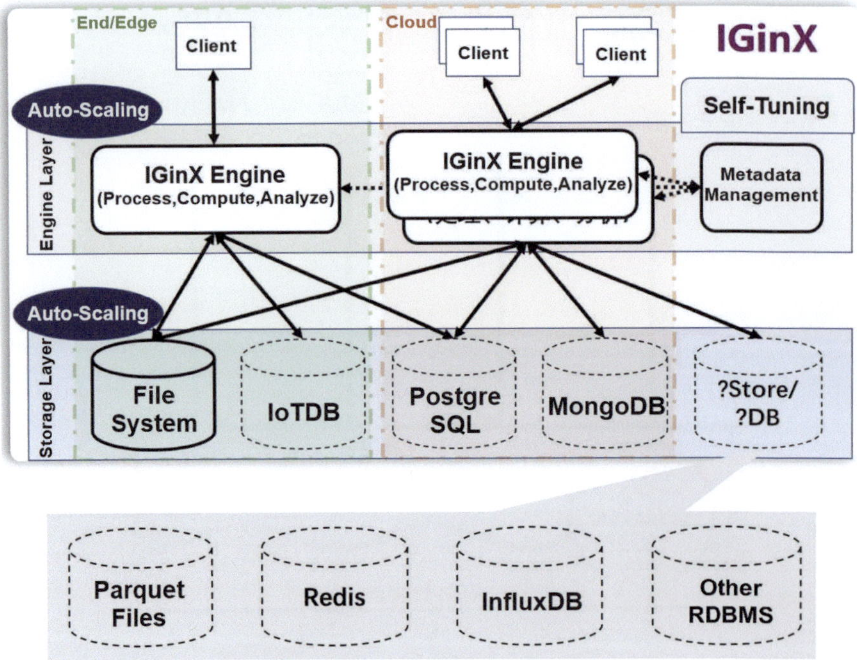

Fig. 10 IGinX's disaggregated storage and computation architecture

aviation engines, from financial risk analysis to intelligent management of medical health records to full lifecycle prediction maintenance and optimal scheduling in intelligent manufacturing and even data-driven planning in smart cities. The integrated data management analysis system is the core of modern digital transformation; it not only reduces the complexity of data management but also opens a new chapter in the realization of data value.

Example: Application of an All-Type Data Management Framework

An aircraft engine manufacturer has developed a cloud-based integrated data management system that can integrate data from multiple stages, such as design, production, testing, and use. Through this platform, engineers and decision-makers can fully understand the operating conditions of each component of the engine and use intelligent algorithms to analyze the data, identify potential problems and take timely repair measures. For example, when a certain engine component exhibits abnormal vibrations, the system can quickly identify and provide repair suggestions, thus avoiding equipment failure due to excessive wear.

Through the application of this platform, designers can optimize the design of new products on the basis of historical performance data, manufacturers can ensure that the engine components produced meet quality standards through the analysis of real-time monitoring data, and the operations team can track the operating status of the engine in real time, predict potential failures and carry out maintenance in advance, thereby reducing downtime and improving operational efficiency.

Framework Value Summary:

- **Improve data utilization efficiency**: Data from all links are integrated, the associations of all-domain data are used, and the value of the data is increased.
- **Optimize decision support**: The integration of all types of massive data, as well as deep analysis and intelligent algorithm support, can help various departments make accurate decisions.
- **Enhanced cross-departmental collaboration**: Comprehensive data management can achieve efficient data sharing and collaboration among design, manufacturing, operations and other departments.

Intelligent Decision Support Across the Full Data Lifecycle

The lifecycle management of aircraft engines covers every stage, from design, manufacturing, installation, operation, and maintenance to retirement. Each stage generates a large amount of data, and once these data are effectively integrated and analyzed, they can provide intelligent support for decision-making at each stage [16, 17]. Traditional maintenance models rely on experience or regular inspections, but this often leads to overmaintenance or delayed repairs, thus increasing unnecessary costs and risks. However, through full lifecycle data management, engine manufacturers and operators can identify potential fault issues in advance on the basis of historical performance data, usage environment data, and real-time operating data to make intelligent decisions and take preventive maintenance measures.

The design and manufacturing stage of the aviation engine is the starting point of full lifecycle data management. At this stage, the design team and engineers generate large amounts of design data, simulation test data, and material performance data. By integrating these data and combining advanced simulation technology, engine manufacturers can optimize design schemes, identify design defects in advance, and improve process flows. Intelligent decision support systems can discover design optimization spaces through deep learning of design data and propose possible improvements. For example, with the help of advanced simulation technology, a design team can predict the performance of an engine under different working environments, thereby optimizing the structure and materials of the engine and improving its performance and reliability. This intelligent data support can greatly increase a product's innovation ability and reduce the maintenance costs associated with unreasonable design.

In the operation and maintenance phase of an aviation engine, the role of an intelligent decision support system becomes increasingly important [18, 19]. By analyzing real-time monitoring data, flight status data, environmental factors, and fault records of the engine during operation, maintenance personnel can achieve accurate fault prediction and risk assessment. Traditional maintenance models often rely on judgment on the basis of experience, which makes discovering potential problems in a timely manner difficult, whereas intelligent analysis of full lifecycle data can provide real-time, accurate fault warnings. For example, by tracking and analyzing the temperature, pressure, vibration, and other data for various parts of the engine for a long time, an intelligent system can identify possible abnormal changes in the parts and propose maintenance suggestions in advance. By implementing this data-based predictive maintenance, not only can the downtime caused by faults be effectively reduced but also the maintenance costs can be significantly reduced, and the overall operating efficiency can be improved.

Intelligent decision support for full lifecycle data can also provide an important reference for the retirement and reuse of aviation engines [20]. At the retirement stage of an engine, relevant data can help manufacturers and operators assess its remaining service life, maintenance history, and performance status to decide whether to remanufacture or reuse it. For example, for an engine that has reached its predetermined service life but still has high residual value, an intelligent decision support system can help assess whether it is suitable for continued use or remanufacturing. Through continuous analysis of full lifecycle data, companies can not only achieve optimal resource allocation but also contribute to environmental protection and sustainable development. This data-driven full lifecycle management model not only improves the efficiency of aviation engine use but also provides a greener and more efficient operating model for the industry.

Example: Intelligent Decision Support

On several Boeing 787 aircraft operated by a certain airline, high-precision sensors and monitoring systems are deployed to collect various operating data of the engine, such as airflow speed, temperature, pressure, and vibration, in real time. These data are transmitted to the maintenance management system at the headquarters through a cloud platform and analyzed in conjunction with historical data and maintenance records. A system analysis revealed that the airflow speed of one of the engines gradually deviated from the normal range, which suggested that there might be a turbine blade wear fault. On the basis of this prediction, the airline arranged maintenance in advance, thus avoiding a serious accident caused by turbine blade faults and ensuring the safety and punctuality of the flight.

Advantages of intelligent decision-making:

- **Preventive Maintenance**: By performing data analysis, the system can identify potential problems in advance to help companies implement preventive maintenance and avoid accidental failures.
- **Cost Optimization**: Through precise maintenance decisions, the cost of over-maintenance is reduced, and the allocation of maintenance resources is optimized.

Fig. 11 Engineers using a maintenance management system to maintain an engine. Sourced from the internet

- **Improved Safety**: Through real-time monitoring and intelligent prediction, the safety and reliability of aviation engines can be maximized (Fig. 11).

Comprehensive Data Management: Enabling Digital Intelligence Transformation

"Digital transformation" refers to a shift from traditional management models to data-driven management models through comprehensive data and intelligent means. In the aviation engine industry, promoting "digital transformation" requires the integration and management of a large amount of heterogeneous data and driving decision-making through intelligent analysis. This transformation not only helps improve work efficiency within an enterprise but also enhances innovation capabilities, safety, and sustainable development capabilities.

First, "digital transformation" requires the effective management of all types of data. This includes multidimensional data, such as design drawings, process parameters, real-time monitoring data, and quality inspection reports, from various links, such as design, manufacturing, testing, and operation, in the product life cycle. These data, such as text, images, and videos, may be structured or unstructured. To effectively integrate and mine data, enterprises need to build a unified data platform and adopt advanced data warehouse technology and a cloud computing architecture so that data from different sources can be integrated and analyzed on the same platform.

Second, the core goal of data management is to enhance decision-making support capabilities through intelligent analysis. In the aviation engine industry, traditional decision-making often relies on experience and human judgment, whereas "digital transformation" can extract potential rules and trends from massive amounts of data by big data analysis, machine learning, and artificial intelligence technologies, thus providing enterprises with a more scientific and accurate decision-making basis. For example, by real-time monitoring of engine operation data and comparative analysis of historical data, potential fault risks can be identified in advance, and predictive maintenance of equipment can be carried out, thereby extending the service life and reducing maintenance costs.

Finally, the management of all types of data not only improves the operational efficiency of an enterprise but also promotes innovation and sustainable development. In the field of aviation engines, precise data analysis and optimization can effectively improve product performance and quality, reduce energy consumption and emissions, and accelerate the research and development cycles of new products. Through in-depth data mining and intelligent design, enterprises can maintain a competitive advantage in technological innovation, respond to environmental regulations and market demands, and promote sustainable green development. Therefore, digital transformation is not only a means for improving enterprise operational efficiency but also the key to the foothold and development of enterprises in the future market.

Example: Practical Application of Digital Transformation

A large aviation engine manufacturer has completed a preliminary layout of "digital transformation" by promoting lifecycle data management. The company not only uses big data and artificial intelligence to optimize products in the design and manufacturing stages but also applies intelligent health management systems in the operation stage. By continuously monitoring the operating status of each engine, along with operational data and maintenance records, the systems can automatically diagnose the health statuses of the engines, discover potential faults in a timely manner and generate maintenance suggestions, thereby greatly reducing the need for manual intervention, lowering maintenance costs, and improving operational efficiency [21].

The advantages of digital transformation:

- **Improved decision-making efficiency**: Data-driven decision support systems can quickly analyze large amounts of data and provide accurate suggestions for decision-makers.
- **Enhanced safety**: Through intelligent monitoring and predictive analysis, the safety and reliability of an engine can be maximized.
- **Reduced costs**: By optimizing decisions in the design, production, operation, and other links, production costs, operation costs, and maintenance costs are reduced.

Case Study: Application of Comprehensive Data Management in an Engine Health Management System

With the continuous advancement of aviation engine technology and the increase in aviation safety requirements, engine health management (EHM) systems have become key tools for airlines and engine manufacturers to ensure the safe and efficient operation of engines. Through all types of data management, EHM systems can not only monitor the operating status of engines in real time but also provide precise predictive maintenance on the basis of the analysis of various data sources, thus maximizing the reliability and maintenance efficiency of engine operation [22, 23].

A leading international airline cooperated with an engine manufacturer to implement an aviation engine health management system on the basis of comprehensive data management. The system integrates data from various stages, such as design, manufacturing, testing, operation, and maintenance, and provides real-time engine status monitoring, health diagnosis, predictive maintenance, and fault warning for airlines through intelligent analysis and decision support (Fig. 12).

Background

The airline operates hundreds of commercial aircraft equipped with various types of aviation engines. The usage data for these engines come from a wide range of sources and include sensor data collected within the engine, in-flight data, daily inspection records, maintenance records, environmental condition data, and historical failure data. However, traditional data management methods cannot effectively

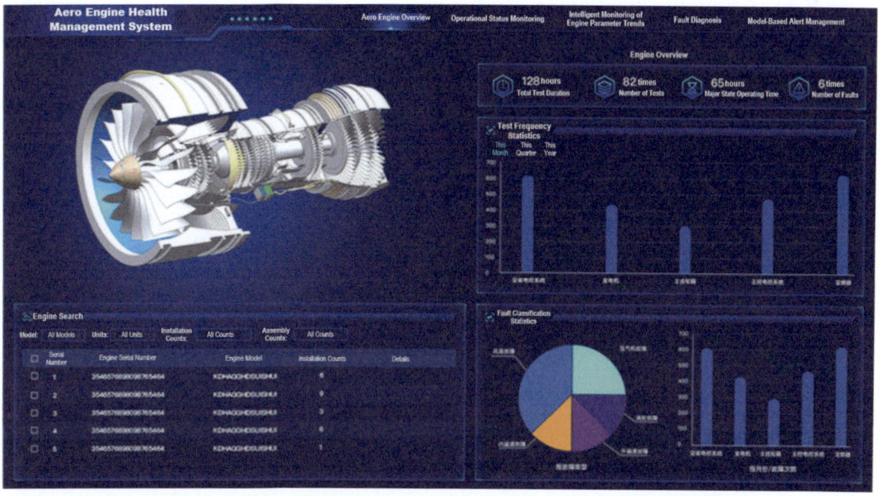

Fig. 12 Aviation engine health management system. Sourced from the internet

integrate or process these heterogeneous data, which leads to information silos and decision-making delays, and cannot achieve real-time, accurate health monitoring or maintenance decision-making.

To address this challenge, the airline decided to introduce an comprehensive data management aviation engine health management system to integrate all types of data and provide analysis and decision support through intelligent algorithms.

Application of Comprehensive Data Management

1. **Data Integration and Cleaning**: The primary task of the aviation engine health management system is to integrate massive amounts of data from multiple sources. These data include sensor data for the engine, flight data, maintenance history, environmental data, operating manuals, and various technical documents from the manufacturing stage. Since these data come from different systems and platforms and the formats and structures are inconsistent, data integration and cleaning constitute the first step in implementing comprehensive data management.

   To overcome this challenge, the airline cooperated with engine manufacturers to develop an efficient data integration platform. Through standardized data interfaces and data conversion mechanisms, the platform achieves automated data access and cleaning. The platform can automatically identify and process data from different sources and can ensure the accuracy, consistency, and completeness of the data, thereby providing a reliable data foundation for subsequent analysis.
2. **Real-Time Monitoring and Health Status Assessment**: During engine operation, the health management system collects various sensor data on the engine in real time, such as temperature, pressure, speed, and vibration data, and compares these data with design parameters and historical data. The system uses intelligent algorithms to assess the health status of the engine and identify possible anomalies. When a piece of data exceeds the set safety threshold, the system immediately issues an alarm and feeds back detailed information about the problem to the operations team.

   For example, during a flight mission, the cylinder temperature of an engine abnormally increased, and the health management system detected this change through real-time monitoring data. The system immediately issued a warning to the ground maintenance personnel and provided an analysis of the cause of the temperature exceeding the standard, which suggested that the operator check the airflow system. In the end, a minor fault in the airflow system was identified through inspection, and timely maintenance prevented potential major faults.
3. **Predictive Maintenance and Intelligent Diagnosis**: By analyzing a large amount of historical engine data, the EHM system can not only monitor the operating condition of the engine in real time but also carry out predictive maintenance based on data analysis. The system combines operating data, environmental

data, and maintenance history and uses machine learning and deep learning algorithms to predict the remaining useful life (RUL) of various parts of the engine. These predictions can help airlines plan maintenance in advance, avoid unplanned downtime and maintenance, and reduce operating costs.

For example, the turbine blades of an engine wear out faster because they are in a high-temperature and high-pressure environment for a long period of time. By analyzing past flight data and maintenance records, the EHM system predicted that these components would reach the critical point of their service life after the next flight and suggested that the airline arrange maintenance in advance. Thus, the airline could address potential faults in advance without affecting flight safety or punctuality, thereby improving the reliability and operational efficiency of the engine.

4. **Fault Analysis and Root Cause Tracing**: Engine failures are often the result of multiple factors working together, and simple fault diagnosis may have difficulty finding the root cause of a problem. Comprehensive data management integrates data from multiple dimensions to provide a more comprehensive perspective for fault analysis. The system can not only identify abnormalities in the engine but also conduct in-depth analysis through historical data and simulation models to find the root cause of a fault.

   For example, in a fault event, the health management system detected a minor leak in an engine's fuel pipe by analyzing the changing trends of multiple data sources (such as vibration, temperature, and speed). Further analysis revealed that the root cause of the leakage was that the fuel pipe did not fully comply with the standard during production, which resulted in microcracks during long-term use. Through this root cause analysis, the airline traced the problem back to the production link and proposed improvement plans to avoid the recurrence of similar problems.

5. **Data Visualization and Decision Support**: To help operators, maintenance engineers, and senior managers understand and use massive amounts of data, the aviation engine health management system provides data visualization functions. The system displays key engine status information, predictive data, and maintenance suggestions in the form of charts, dashboards, etc., so that relevant personnel can intuitively understand the health status of the engine and its maintenance needs.

   Through data visualization, airline management can understand the statuses of different engines in real time and make more scientific dispatch decisions. For example, if an engine is about to reach its scheduled maintenance cycle, the system will prompt management through a visual interface to arrange maintenance in advance to optimize flight arrangements and avoid operational interruptions caused by failures (Fig. 13).

Fig. 13 Typical closed-loop process of full-type data management applied in the field of aviation engines

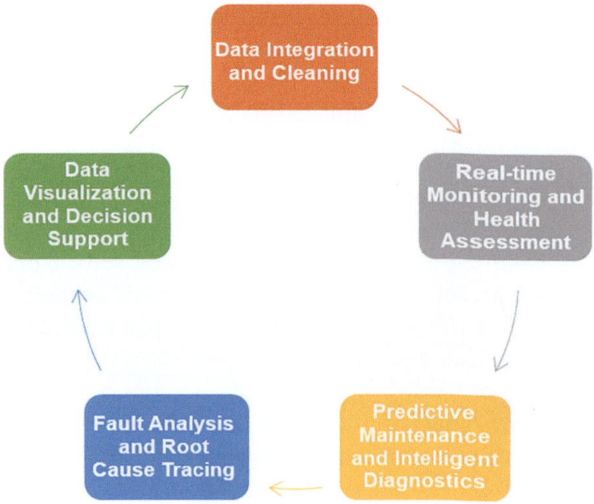

Achievements and Advantages

1. **Enhancing operational safety**: A full-type data management system effectively reduces the occurrence of sudden events caused by faults and equipment aging through real-time monitoring and intelligent analysis. Through predictive maintenance, airlines can identify potential problems in advance and take measures in a timely manner to avoid the effect of sudden failures on flight safety.
2. **Reducing maintenance costs**: Through intelligent health management and predictive maintenance, airlines can optimize maintenance plans and avoid unnecessary regular inspections and excessive maintenance. This greatly reduces maintenance costs and improves resource utilization efficiency. Airlines can also reduce the frequency of manual inspections and maintenance through intelligent systems, thus further reducing costs.
3. **Enhancing operational efficiency**: With the full visualization of engine health status and intelligent decision support, airlines can better arrange maintenance and dispatch, reduce engine downtime, and improve flight punctuality and overall operational efficiency.
4. **Promoting continuous improvement**: The health management system is not only a monitoring tool but also a data analysis platform. By accumulating and analyzing the operating data of an engine's entire lifecycle, airlines can provide data support for subsequent engine design and production to continuously optimize product quality and design schemes.

Future Prospects

With the continuous development of data technology and AI algorithms, aviation engine health management systems will further deepen the integration and analysis of all types of data, thus promoting more intelligent decision support and maintenance management. In the future, health management systems will not only be able to more accurately predict the health statuses of various engine components but also achieve cross-system data collaboration and automated decision-making through deep integration with the airline's operational dispatch system, flight planning system, etc.

In addition, with the development of technologies such as 5G and the Internet of Things (IoT), the real-time coverage of aviation engine health management systems will further expand. Through the deployment of more sensors and the transmission of real-time data, airlines can more accurately grasp the operating status of each engine, thus providing strong technical support for the realization of intelligent aviation on a global scale.

In summary, the application of full-type data management in aviation engine health management systems not only improves operational safety and reduces maintenance costs but also greatly enhances the efficiency and reliability of flight operations. Against the backdrop of the aviation industry's increasing emphasis on intelligent and digital transformation, this system provides important technical support for promoting the aviation industry toward a digital, intelligent, and green future.

Data- and AI-Driven Future Innovations

With the continuous advancements in technology, artificial intelligence (AI) and big data technology are gradually changing the design, manufacturing, operation, and maintenance processes of aviation engines. The combination of AI and big data can not only optimize existing technologies but also lay a solid foundation for the innovation of the next generation of engines. In the future, AI will play a greater role in all aspects of engines, promote rapid development in terms of intelligence, greenness, and efficiency, and become the core driving force for future innovation in the aviation industry.

Prospective Roles of AI in Engine Innovation

In the future development of aviation engines, artificial intelligence (AI) will play an increasingly important role, especially in the design and optimization of engines. AI can not only help designers optimize multiple dimensions through intelligent algorithms in the early stages of R&D but also provide more accurate intelligent decision support in the operation and maintenance of engines. The advantage of AI lies in its powerful data processing ability; it can carry out deep learning on the basis of large amounts of historical data and real-time data to identify potential design improvement points and optimization schemes.

With the rapid development of AI technologies such as deep learning, reinforcement learning, and natural language processing, AI can learn from complex historical data, perform pattern recognition and optimization, and find the most suitable design or operation strategy for an engine. This makes the design and operation of the engine more efficient, energy-saving, and intelligent, thus promoting further innovation in engine technology. For example, AI can simulate different working environments and conditions, test and evaluate the performance of the engine, and thus predict its performance in actual applications. AI can adjust the design scheme in real time, avoid errors in the design, and discover potential problems, thereby making the R&D process for engine products more efficient and accurate.

Example: AI Driving Innovation

An aircraft engine research and development company uses AI technology for the innovative design of turbine blades. The AI system, by analyzing massive amounts of simulation data, automatically identifies the optimal blade geometry, thereby optimizing the aerodynamic performance and fuel efficiency of the blade. Compared with traditional design methods, the application of AI not only improves fuel efficiency but also reduces noise pollution during engine operation. In addition, AI plays a significant role in engine fault diagnosis and prediction. By analyzing engine operating data under different usage conditions, AI can predict potential faults in advance and provide precise repair suggestions to maintenance personnel, thereby avoiding sudden equipment failures and extending the life of the engine.

Summary of Prospective Roles:

- AI helps designers perform efficient multidimensional optimization, thus reducing trial and error costs.
- AI enhances the intelligence level of engines, thereby promoting more efficient and environmentally friendly design.
- AI can play a significant role during engine operation by predicting potential faults in advance and reducing maintenance costs.

Promoting Sustainable Development and Green Technology

With the increasing global demand for environmental protection, the aviation industry faces unprecedented environmental pressures. Reducing energy consumption, improving fuel efficiency, and reducing pollution emissions have become key goals for aircraft engine research and development. Big data and AI technology not only help improve engine performance but also provide strong support for the green transformation of aircraft engines. Through data analysis and intelligent optimization, AI can facilitate the design of more efficient and environmentally friendly engines and promote breakthroughs in green technology in the industry.

The combination of AI and big data technology can help engineers optimize the combustion process of engines, improve the design of turbine blades, and optimize the structure of the combustion chamber, thereby significantly improving combustion efficiency and reducing harmful gas emissions. For example, AI can be used to simulate and optimize different combustion modes, find the best combustion chamber design, improve combustion efficiency, and reduce carbon emissions. Moreover, AI can also help airlines optimize flight route planning through real-time analysis of flight data to reduce fuel consumption and emissions and achieve green operation goals.

In addition, AI and big data can help airlines predict the fuel consumption of engines on the basis of real-time data and adjust the combustion process according to actual needs, thereby improving energy utilization while reducing pollutant emissions. These technologies can help airlines not only achieve green transformation but also respond to environmental policies in the context of global carbon reduction, thus enhancing the industry's sustainable development capabilities.

Example: Green Technology Application

An aircraft engine company, by introducing AI and data analysis technology, successfully optimized the combustion process of an engine. Through simulation and intelligent optimization algorithms, a more efficient combustion chamber structure was designed to minimize fuel consumption and harmful gas emissions. This technology not only made significant breakthroughs in environmental protection but also effectively improved the fuel efficiency of the engine, thus reducing the fuel costs of airlines. In addition, AI helps airlines analyze flight route data and flight modes to further reduce carbon emissions in operation through precise prediction and planning. The application of these green technologies not only helps airlines save costs but also contributes to the sustainable development of the global aviation industry.

Summary of the Green Transformation:

- AI and data analysis have been used to optimize the combustion process, improve fuel efficiency, and reduce carbon emissions.
- AI has promoted the development of green technology to enable airlines to respond actively to global environmental policies while reducing operating costs.
- The application of green technology has not only improved the environmental image of airlines but also enhanced the industry's sustainable development capabilities.

Future Intelligent Engines: Green, Intelligent, and Efficient

Future aircraft engines will not only focus on mechanical performance and reliability but also emphasize overall environmental protection, intelligence, and high efficiency. With the rapid development of AI, IoT, and sensor technology, future engines will have more intelligent self-diagnosis, optimization, and adjustment capabilities and will even be able to adapt automatically on the basis of the flight environment and status.

In the future, intelligent engines will use more built-in sensors and AI algorithms that are capable of real-time monitoring of various engine operating data, including combustion efficiency, temperature, pressure, and vibration data. These sensors will not only collect data in real time but also link with external environmental data, flight data, and weather conditions, thereby automatically optimizing engine operation on the basis of the current working status. The engines will automatically adjust the combustion process during flight, optimize fuel efficiency and reduce emissions, thus ensuring efficient, environmentally friendly, and safe operation.

In addition, intelligent engines will also be equipped with self-diagnosis and maintenance systems that are capable of real-time fault identification and self-repair or automatic adjustment. These highly intelligent self-maintenance systems will make the engines more reliable, reduce the need for human intervention, and lower maintenance costs. Through a combination of AI and IoT technology, the engines will communicate with the ground control center in real time during flight, provide more accurate operation status feedback, and adjust flight operations on the basis of system analysis results.

Example: Future Intelligent Engine Development

An airline has collaborated with an engine manufacturer to develop an intelligent engine on the basis of AI and IoT technology. This engine is equipped with multiple built-in sensors, which monitor combustion efficiency, air temperature, vibration, etc., in real time and are optimized through the built-in AI system. The engine can automatically adjust the combustion process according to the flight status to maximize fuel efficiency and in the event of any abnormalities, thus ensuring that the engine can maintain efficient, safe, and low-emission operation under any conditions. The launch of this intelligent engine not only improves the operational efficiency of the airplane during flight but also provides strong technical support for the green transformation of the future aviation transportation industry.

Future Prospects of Intelligent Engines:

- **Green Operation and Environmental Protection**: Future intelligent engines will further reduce carbon emissions and environmental pollution by optimizing the combustion process and automatically adjusting fuel consumption.

- **Intelligent Decision-making**: Through a built-in AI system, intelligent engines can automatically self-diagnose and adjust, reduce the failure rate, and improve safety.
- **Efficient Operation**: Future engines will optimize operation through sensor measurements and AI analysis, thus improving fuel efficiency and cost-effectiveness.

Conclusions

With the continuous advancements in data and intelligent technologies, the aviation engine industry is undergoing profound changes. From design to manufacturing and from operation to maintenance, data-driven and intelligent technology are driving the innovation and development of aviation engines. Facing the arrival of the digital age, aviation companies must seize opportunities, meet challenges, actively promote technological innovation, enhance digital capabilities, optimize intelligent decision-making systems, and prepare for future market competition.

This chapter discusses the integrated application of AI and big data in detail, especially in design optimization, intelligent manufacturing, and full lifecycle management, and emphasizes the importance of integrated data management for the transformation of the aviation engine industry toward "digital intelligence". The integration of full lifecycle data and intelligent decision support is the core strategy for future innovation in the aviation engine industry, and strong data management capabilities provide an indispensable foundation for this process. These technologies not only improve the efficiency and reliability of existing engines but also lay a foundation for future green, intelligent, and efficient engines.

References

1. Bisanti GM, Mainetti L, Montanaro T, Patrono L, Sergi I (2023) Digital twins for aircraft maintenance and operation: A systematic literature review and an IoT-enabled modular architecture. Internet Things 24:100991
2. Xiong M, Wang H, Qiang F, Yi X (2021) Digital twin–driven aero-engine intelligent predictive maintenance. Int J Adv Manuf Technol 114(11):3751–3761
3. Madrid J (2018) Design for manufacturing and producibility in fabricated aerospace structures. Chalmers Tekniska Hogskola (Sweden)
4. Forest F, Lacaille J, Lebbah M, Azzag H (2018) A generic and scalable pipeline for large-scale analytics of continuous aircraft engine data. In: 2018 IEEE international conference on big data (big data). IEEE, pp 1918–1924
5. Wang W, Ma J (2024) A review: applications of machine learning and deep learning in aerospace engineering and aero-engine engineering. Adv Eng Innov 6:54–72
6. Guo ZD, Hui C, Chen Y, Jiang SM, Song LM, Li J, Feng ZP (2023) Research on the prediction method of turbine blade flow field based on the principle of similarity. Chin J Mech 55(11):2647–2660

7. Ghosh S, Padmanabha GA, Peng C, Andreoli V, Atkinson S, Pandita P, Vandeputte T, Zabaras N, Wang L (2022) Inverse aerodynamic design of gas turbine blades using probabilistic machine learning. J Mech Des 144(2):021706
8. Simon DL, Garg S, Hunter GW, Guo T-H, Semega KJ (2004) Sensor needs for control and health management of intelligent aircraft engines. Turbo Expo Power Land Sea Air 41677:873–882
9. Li J, Zhao G, Zhang P, Maocheng X, Cheng H, Pengfei H (2023) A digital twin-based on-site quality assessment method for aero-engine assembly. J Manuf Syst 71:565–580
10. Huang Y, Tao J, Sun G, Tengyun W, Liling Y, Zhao X (2023) A novel digital twin approach based on deep multimodal information fusion for aero-engine fault diagnosis. Energy 270:126894
11. Chetan M, Yao S, Griffith DT (2021) Multi-fidelity digital twin structural model for a sub-scale downwind wind turbine rotor blade. Wind Energy 24(12):1368–1387
12. Feng G, Cui D, Wang C, Jiapeng Y (2009) Integrated data management in complex product collaborative design. Comput Ind 60(1):48–63
13. Armbrust M, Ghodsi A, Xin R, Zaharia M (2021) Lakehouse: a new generation of open platforms that unify data warehousing and advanced analytics. Proc CIDR 8:28
14. Zhu Y et al (2025) Iginx. https://github.com/IGinX-THU/IGinX
15. Zhang LZ, Zhu YQ, An YZ, Zi Y, Wang JM (2024) The optimization problem of system load balancing in industrial Internet of Things data management and its solution. Sci China Inf Sci 54(10):2343–2367
16. Gogineni S, Exner K, Stark R, Nickel J, Oeler M, Witte H (2019) Semantic assistance system for providing smart services and reasoning in aero-engine manufacturing. In: Metadata and semantic research: 13th international conference, MTSR 2019, Rome, Italy. Revised selected papers. Springer, pp 90–102
17. Li J, Zhou G, Zhang C (2022) A twin data and knowledge-driven intelligent process planning framework of aviation parts. Int J Prod Res 60(17):5217–5234
18. Mustapha F, Sapuan SM, Ismail N, Mokhtar AS (2004) A computer-based intelligent system for fault diagnosis of an aircraft engine. Eng Comput 21(1):78–90
19. Patel SA, Kamrani AK (1996) Intelligent decision support system for diagnosis and maintenance of automated systems. Comput Ind Eng 30(2):297–319
20. Zhang P, Zhao S-W, Tan B, Yu L-M, Hua K-Q (2011) Applications of decision support system in aviation maintenance. In: Efficient decision support systems-practice and challenges in multidisciplinary domains–InTech, pp 397–412
21. Zhao J, Li Y-G, Sampath S (2023) A hierarchical structure built on physical and data-based information for intelligent aero-engine gas path diagnostics. Appl Energy 332:120520
22. Volponi AJ (2014) Gas turbine engine health management: past, present, and future trends. J Eng Gas Turbines Power 136(5):051201
23. Wang Z, Wang Y, Wang X, Zhao B, Zhao Y (2024) A novel approach to aeroengine performance diagnosis based on physical model coupling data-driven using low-rank multimodal fusion method. J Glob Power Propuls Soc 8:334–348

MIX
Papier aus verantwortungsvollen Quellen
Paper from responsible sources
FSC® C105338

If you have any concerns about our products,
you can contact us on
ProductSafety@springernature.com

In case Publisher is established outside the EU,
the EU authorized representative is:
Springer Nature Customer Service Center GmbH
Europaplatz 3, 69115 Heidelberg, Germany

Printed by Libri Plureos GmbH
in Hamburg, Germany